T0136310

The GETMe Mesh Smoothing Framework

A Geometric Way to Quality Finite Element Meshes

The GETMe Mesh Smoothing Framework

A Geometric Way to Quality Finite Element Meshes

Dimitris Vartziotis
Joachim Wipper

CRC Press
Taylor & Francis Group
Boca Raton London New York

CRC Press is an imprint of the
Taylor & Francis Group, an **informa** business

CRC Press
Taylor & Francis Group
6000 Broken Sound Parkway NW, Suite 300
Boca Raton, FL 33487-2742

© 2019 by Taylor & Francis Group, LLC
CRC Press is an imprint of Taylor & Francis Group, an Informa business

---

Library of Congress Cataloging-in-Publication Data

Names: Vartziotis, Dimitris, author. | Wipper, Joachim, author.
Title: The GETMe mesh smoothing framework : a geometric way to quality finite element meshes / Dimitris Vartziotis and Joachim Wipper.
Other titles: Geometric element transformation method mesh smoothing framework
Description: Boca Raton : Taylor & Francis Group, 2018. | Includes bibliographical references and index.
Identifiers: LCCN 2018040529 | ISBN 9780367023423 (hardback : alk. paper)
Subjects: LCSH: Finite element method. | Subdivision surfaces (Geometry) | Geometry--Data processing.
Classification: LCC QC20.7.F56 V368 2018 | DDC 518/.25--dc23
LC record available at https://lccn.loc.gov/2018040529

Visit the Taylor & Francis Web site at
http://www.taylorandfrancis.com

and the CRC Press Web site at
http://www.crcpress.com

Contents

1

Introduction

Meshes are gapless and overlapping free tessellations of domains, such as surfaces or volumes, into primitives such as triangles, quadrilaterals, tetrahedra, or hexahedra. Using geometrically simple elements of only one or a few preferably polytope types brings both the flexibility to represent large-scale domains by locally placed small elements and the flexibility to represent complex domains by uniformly shaped simple elements.

Areas of application range from art and architecture to the general field of approximation theory, including mathematical, physical, engineering, and design applications. For example, Figure 1.1 depicts a detail of the Medusa floor mosaic found in Zea (Piraeus) dating back to the 2nd century AD. A paragon for meshes used in design and simulation is the Munich Olympic Park roof shown in Figure 1.2, a tensile structure of acrylic panels constructed by Frei Otto and Günther Behnisch. This roof, representing the Alps, is also a milestone with respect to physical simulation techniques, namely the finite element method used for design, development, and analysis of the tensile structure by John Argyris and Fritz Leonhardt. Similarly, Figure 1.3 depicts the Aletis car model and mesh by TWT GmbH Science & Innovation combining design and simulation aspects.

These different fields of application often come with their own specific mesh quality requirements; for example, they may be aesthetically motivated in design and computer visualization applications or mathematically motivated in the case of finite element methods. In both cases, meshes usually have to be conformal and of good quality. That is, two elements are either disjoint or share one common node, edge or face, respectively, and each element must fulfill certain quality criteria with respect to its shape. In this context, regularly shaped elements are often preferred.

Ideally, mesh quality is an objective of the mesh generation process. For simple domains, this can, for example, be ensured by using parametric generation processes for structured meshes. A more flexible and generally applicable meshing method is the Delaunay triangulation. In a Delaunay mesh, each circumcircle of a triangle contains no other node of the mesh. It follows that the Delaunay triangulation of a set of nodes maximizes the smallest interior angle over all triangles. Delaunay triangulation is fast, flexible, robust, and generally applicable for simplicial mesh generation in arbitrary dimensions. Furthermore, it generates meshes of reasonable quality.

In contrast, mesh generation involving non-simplicial polytopes such as

Figure 1.1: Detail of the Medusa floor mosaic found in Zea (Piraeus), 2nd century AD. Image provided by user Jebulon, Wikimedia Commons, license CC0 1.0 [77].

hexahedra as elements becomes significantly more complex, since the conformal mesh criterion imposes strong structural restrictions for element connectivity. In such cases a common scheme is to partition the domain to be meshed into simple-shaped subdomains, which can be meshed individually by a parameterized approach considering the conformal interface meshes of such subdomains. Depending on the application, more complex elements can have favorable properties. Hence, mesh generation should find a compromise between computational and implementational complexity, flexibility, and the resulting mesh quality.

Therefore, it is common practice to address mesh quality aspects by a separate improvement step after mesh generation. This also holds for applications where the simulation domain changes its shape over time. To avoid a costly generation of a new mesh within each simulation step, the existing mesh is transformed and improved. This can either be done by local topological modifications, like splitting, inserting, or removing elements, or by node relocations only. The latter mesh improvement methods are called smoothing methods.

Depending on the element type, a local topological modification might propagate through the entire mesh to preserve conformity. For example, in a regular three-dimensional grid, splitting a cube into two equal hexahedra leads to a splitting of the mesh along an entire plane. Hence, in practice, mesh smoothing methods are preferred if the quality problems are not caused by adverse topological configurations.

One of the most common and simplest mesh smoothing methods is Laplacian smoothing, where each node is replaced by the arithmetic mean of its

Figure 1.2: Tensile structure in the Munich Olympic Park. Image provided by Diego Delso, delso.photo, Wikimedia Commons, license CC-BY-SA [32].

edge-connected neighbors. This smoothing scheme is iteratively applied a given number of times to all inner mesh nodes or iteratively applied until node positions converge. Due to its implementational simplicity and computational efficiency, Laplacian smoothing is still popular, although it might lead to the generation of inverted elements. These are elements that violate a prescribed orientation or have a negative determinant. Smart Laplacian smoothing methods avoid the generation of inverted elements by node movement restrictions or node resetting techniques. Laplacian smoothing belongs to the class of geometric smoothing schemes, since its node relocation approach and its termination criterion are entirely based on geometric entities.

Mesh quality is a matter of the quality criterion under consideration. The choice of the quality criterion itself depends on the field of application. Having a specific quality criterion in mind, it is an obvious approach to incorporate it into an objective function used to improve quality by means of mathematical optimization. Local optimization-based smoothing methods use this approach to move nodes of a single element or nodes shared by some elements to improve local mesh quality. In contrast, global optimization-based mesh smoothing methods incorporate the quality of all mesh elements into one objective function and relocate all non-fixed mesh nodes within one iteration step. Such steps are iterated until quality improvement drops below a given threshold. Using a strict mathematical optimization approach, these methods aim to reach at least a local maximum of mesh quality, thus providing the benchmark with respect to quality. However, using optimization-based methods comes at the cost of higher implementational and computational complexity, due to the

Figure 1.3: Aletis car model and mesh by TWT GmbH Science & Innovation combining design and simulation aspects.

evaluation of quality functions and numerical optimization. Therefore, various hybrid methods have been proposed to find a balance between complexity and quality.

Many element quality criteria are based on measuring the deviation of an arbitrary element from its regular counterpart. They differ, for example, in computational complexity, generalizability with respect to different element types, and applicability with respect to optimization-based methods. Regular elements offer a high degree of symmetry. This is particularly favorable in the context of finite element methods. Here, mesh elements are used as supports for locally defined trial functions, building the basis for function spaces in which a physical solution of a partial differential equation is sought. Thus, the shape of the mesh elements has an impact on the trial functions, and, with this, on the approximation power of the associated function space.

The foundation of the geometric element transformation-based mesh smoothing framework described in this work lies within a geometric play-around with paper and pen. By constructing similar triangles on the sides of a polygon and taking the apices as new polygon points, one of the authors observed that the resulting polygon became more and more regular if applied iteratively [140]. A question emerged as to whether such a transformation could be used for mesh improvement by iteratively transforming the mesh element with the lowest quality, and this set the seed point for developing the geometric element transformation method, abbreviated by the acronym

GETMe. Since these first steps in 2007, the theory of regularizing geometric element transformations and the associated GETMe smoothing algorithms have been developed into a powerful smoothing framework with a wide range of applications.

Regularizing transformations have been derived for polygons with an arbitrary number of nodes as well as for the most relevant volumetric elements. All transformations are based on simple geometric constructions, which increase element regularity if applied iteratively. Two different basic geometric element transformation smoothing schemes have been developed. One is the already described approach of iteratively transforming the element with the lowest quality, the other is based on transforming all elements of the mesh and deriving new mesh node positions as arithmetic or weighted means of transformed element nodes. Whereas the first approach is geared towards improving the worst elements of the mesh, the second method is geared towards improving the entire mesh. These two methods are combined by the GETMe method to improve both mesh quality aspects. Furthermore, an adaptive approach is described, which controls these steps dynamically.

All these GETMe variants have in common that element improvement is incorporated intrinsically by the regularizing element transformation. That is, the incorporation of quality metrics is not required for mesh improvement. Therefore GETMe, like Laplacian smoothing, belongs to the class of geometric mesh transformation methods. Nevertheless, quality metrics can be incorporated for a quality-oriented control of the smoothing process and its parameters. From the first tests onwards, applying GETMe smoothing led to surprisingly high-quality meshes reaching the same quality level as global optimization-based approaches. As it turned out recently, this is based on the fact that there is a link between element transformations and gradient flow-based optimization, which explains the remarkably good results.

This work aims to describe both regularizing element transformations and the smoothing schemes build on these. Giving a brief introduction to the finite element method and providing numerical examples for both mesh smoothing and finite element approximation not only demonstrates the benefits of GETMe-based mesh smoothing with respect to the resulting mesh quality alone, but also with respect to finite element solution accuracy and efficiency. Although finite element mesh smoothing is primarily addressed in this book, GETMe smoothing is applicable to any mesh smoothing context aiming at high-quality meshes. Mathematical as well as implementational aspects are discussed. Therefore, this work addresses both readers interested in the mathematical foundations of regularizing element transformations and those who are interested in implementing or applying GETMe smoothing.

The rest of this book is organized as follows. First, meshes and their elements are introduced in Chapter 2. Quality metrics for elements and meshes are discussed, and a brief overview of mesh generation methods is given. For readers interested in a finite element application context for smoothing methods, Chapter 3 introduces the mathematical foundation of the finite element method with the example of an elliptic boundary value problem, and discusses the influence of mesh quality on solution accuracy and efficiency. The last introductory chapter, Chapter 4, provides an overview of existing mesh improvement methods that are based either on node relocation only or on topological modifications.

Chapter 5 describes regularizing element transformations for polygonal elements with an arbitrary number of nodes based on constructing similar triangles on the sides. It also provides a mathematical proof of the regularizing property of such transformations based on circulant matrices and the discrete Fourier transform. Furthermore, transformations for the most common volumetric elements are given, accompanied by numerical results demonstrating their regularizing effect.

Such transformations are the key ingredient in the geometric element transformation-based smoothing framework described in Chapter 6. This consists of three simple approaches to mesh smoothing by using regularizing transformations, focusing primarily on either improving lowest-quality elements or all elements of the mesh. Here, quality-related and quality-criterion-free variants exist. These building blocks are combined, resulting in the two-phase GETMe smoothing scheme aiming at improving overall mesh quality as well as minimal element quality. It is generalized to an adaptive smoothing scheme where the elements to transform are selected by a variable quality threshold. The resulting smoothing methods are easy to implement and combine the computational efficiency of simple geometry-based smoothing approaches with the high mesh quality results only obtained so far by costly global optimization-based methods. Finally, Chapter 6 discusses various aspects and characteristics of GETMe smoothing.

A broad variety of smoothing examples is given in Chapter 7. These include smoothing of triangular, quadrilateral, and mixed polygonal planar meshes, anisotropic mesh smoothing, and smoothing of surface and volumetric meshes. The latter addresses real-world examples for tetrahedral, hexahedral, and mixed volumetric meshes, including pyramidal and prismatic elements. Furthermore, the impact of mesh smoothing on finite element solution efficiency and accuracy is discussed using the example of Poisson's equation for planar and volumetric simulation domains. This includes efficiency comparisons for the incorporated solver of the associated Ritz-Galerkin system, their condition numbers, and the comparison of analytic solution errors with respect to various norms. Where applicable, the results are also compared to those obtained by smart Laplacian smoothing, as well as a state-of-the-art global optimization-based smoothing approach.

Finally, Chapter 8 describes recent developments in the theory of GETMe smoothing. It extends the framework by providing alternative regularizing element transformations and alternative smoothing approaches following the spirit of GETMe smoothing. The link to gradient flow-based optimization methods is established, and this gives an indication of why GETMe smoothing results in global optimization-like mesh quality.

2

Elements and meshes

In this chapter, definitions for the most common geometric elements used in finite element simulations are provided. They represent the building blocks for meshes tessellating prescribed simulation domains in two- and three-dimensional space. To assess the quality of meshes and their elements, the mean ratio quality number is described and its properties are discussed. In addition, a brief overview of mesh generation methods is given and concepts of Delaunay meshing are exemplified.

2.1 Elements

2.1.1 Elements for planar and surface meshes

The most common elements for planar or surface meshes are triangles and quadrilaterals. These elements are depicted in Figure 2.1 together with a general *polygonal element* with n nodes. Here, the elements are shown in their regular configuration with their node numbering scheme. In preparation for the polygonal element transformations and their analysis given in Chapter 5, the indices of the polygonal element nodes $p_i \in \mathbb{R}^m$, $i \in \{0, \ldots, n-1\}$ are zero-based. Furthermore, polygonal elements are oriented counterclockwise with respect to the normal of a prescribed plane or surface. For planar elements, it holds that $m = 2$, and for surface elements $m = 3$ is considered.

As can be seen, independent of the number of element nodes, each node has two neighboring nodes, each connected by one edge. The general polygonal

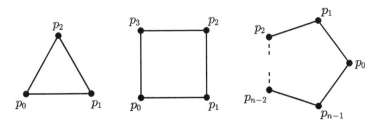

Figure 2.1: Regular polygonal mesh elements.

element with n nodes consists of n edges connecting the nodes p_0 and p_1, p_1 and p_2, and so on. The last edge connects the nodes p_{n-1} and p_0. Using zero-based node indices allows to define the ith edge as a line segment connecting the nodes p_i and p_{i+1} with indices taken modulo n.

The only cases where two edges of a valid polygonal element have common points are two neighboring edges sharing one node. This excludes, in particular, self-intersecting polygons as boundaries of valid polygonal elements. The counterclockwise orientation of the valid element ensures that the interior of the element always lies on the left with respect to its oriented boundary. A precise definition for element validity is given later on by Definition 2.1.

2.1.2 Elements for volumetric meshes

The most common *polyhedral elements* are the tetrahedron, hexahedron, pyramid, and prism depicted in Figure 2.2 from left to right. The figure also shows the node numbering scheme for these elements used throughout the rest of this book. All elements are depicted in their regular form; that is, in their configuration with the highest level of symmetry. A proper definition of element regularity based on the mean ratio quality criterion is given in the following section.

All elements can be written as $3 \times n$ node coordinates matrices, i.e., $E^{\text{tet}} = (p_1, \ldots, p_4)$, $E^{\text{hex}} = (p_1, \ldots, p_8)$, $E^{\text{pyr}} = (p_1, \ldots, p_5)$, and $E^{\text{pri}} = (p_1, \ldots, p_6)$, with column vectors $p_i \in \mathbb{R}^3$ representing the nodes. Node coordinates matrices of regular elements with edge length one are given, for example, by

$$E^{\text{tet}} = \begin{pmatrix} 0 & 1 & 1/2 & 1/2 \\ 0 & 0 & \sqrt{3}/2 & \sqrt{3}/6 \\ 0 & 0 & 0 & \sqrt{2/3} \end{pmatrix},$$

$$E^{\text{hex}} = \begin{pmatrix} 0 & 1 & 1 & 0 & 0 & 1 & 1 & 0 \\ 0 & 0 & 1 & 1 & 0 & 0 & 1 & 1 \\ 0 & 0 & 0 & 0 & 1 & 1 & 1 & 1 \end{pmatrix},$$

$$E^{\text{pyr}} = \begin{pmatrix} 0 & 1 & 1 & 0 & 1/2 \\ 0 & 0 & 1 & 1 & 1/2 \\ 0 & 0 & 0 & 0 & 1/\sqrt{2} \end{pmatrix},$$

$$E^{\text{pri}} = \begin{pmatrix} 0 & 1 & 1/2 & 0 & 1 & 1/2 \\ 0 & 0 & \sqrt{3}/2 & 0 & 0 & \sqrt{3}/2 \\ 0 & 0 & 0 & 1 & 1 & 1 \end{pmatrix}.$$

The boundary of the tetrahedral element consists of four triangular faces, whereas the boundary of the hexahedral element consists of six quadrilateral faces. In contrast, the pyramidal, as well as the prismatic element, have triangular as well as quadrilateral faces. Due to this difference in the number of face nodes, there is no matrix representation for the node indices of the

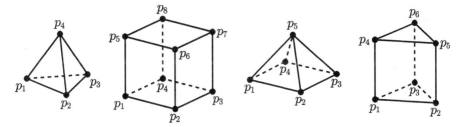

Figure 2.2: Regular polyhedral mesh elements.

faces. Therefore, an alternative representation based on the *face node tuples*

$$F^{\text{tet}} := \big((1,2,3),(1,2,4),(2,3,4),(3,1,4)\big), \tag{2.1}$$

$$F^{\text{hex}} := \big((1,2,3,4),(1,2,6,5),(2,3,7,6),(3,4,8,7),$$
$$(4,1,5,8),(5,8,7,6)\big), \tag{2.2}$$

$$F^{\text{pyr}} := \big((1,2,3,4),(1,2,5),(2,3,5),(3,4,5),(4,1,5)\big), \tag{2.3}$$

$$F^{\text{pri}} := \big((1,2,3),(1,2,5,4),(2,3,6,5),(3,1,4,6),(4,6,5)\big) \tag{2.4}$$

is given. Here, the kth element of a faces tuple $F \in \{F^{\text{tet}}, F^{\text{hex}}, F^{\text{pyr}}, F^{\text{pri}}\}$ represents the tuple of node indices defining the oriented kth face. The index of the ith node of the kth face will be denoted as $F_{k,i}$. For example, the second face of the pyramid is the triangle defined by the nodes p_1, p_2, and p_5, and the fourth face of the prism is the quadrilateral defined by the nodes p_3, p_1, p_4, and p_6.

As mentioned before, in contrast to the tetrahedron and the hexahedron, the face tuples of the pyramid and the prism differ in their number of elements. That is, the pyramid and the prism have faces of different types. Furthermore, in contrast to all other element nodes having three edge-connected neighbors, the apex node of the pyramidal element has four edge-connected neighbors.

In the case of the triangular element in two dimensions, and the tetrahedral element in three dimensions, each node is connected to all other nodes of the element. In general, elements with n nodes in $\mathbb{R}^n$ with such a complete topological connectivity are called *simplicial elements*. They are of particular interest, since they offer a great flexibility in meshing. Furthermore, efficient meshing algorithms based on Delaunay tessellation exist, as will be described in Section 2.4.2.

2.2   Meshes

2.2.1   Introduction

Let Ω denote an ℓ-dimensional domain embedded into $\mathbb{R}^m$, where $\ell \leq m$. In the following, a *mesh* M denotes a gapless decomposition or approximation of Ω into ℓ-dimensional polytopes. The polytopes of the mesh are called *elements*. Their boundary is in turn hierarchically assembled of k-polytopes with $k \in \{0, \ldots, \ell-1\}$. For example, the boundary of the volumetric tetrahedral element is assembled of four triangular faces, sharing common edges, which in turn connect common nodes.

A mesh is called *conformal* if two different ℓ-polytopes are either disjoint or share one common k-polytope with $k < \ell$. The most relevant polytopes are given in Table 2.1. For planar meshes, i.e., $\ell = m = 2$, this means that a planar domain is decomposed into polygonal elements. Different elements either have no common point, share one common node ($k = 0$), or have one common edge ($k = 1$). Volumetric meshes, i.e., $\ell = m = 3$, consist of polyhedral elements that are either pairwise disjoint, or that share one common node ($k = 0$), one common edge ($k = 1$), or one common face ($k = 2$). A mesh with $\ell \neq m$ is, for example, a surface mesh ($\ell = 2$) in three-dimensional space ($m = 3$).

Table 2.1: Polytopes for 2D and 3D meshing.

Dimension k	Name	Example
0	node	•
1	edge	•——————•
2	polygon (face)	△
3	polyhedron (volume)	◁

Meshes with nodes being located inside an edge or face of another mesh element are denoted as *non-conformal*. Such nodes are usually called *hanging nodes*. However, since in a finite element simulation context conformal meshes are preferred, non-conformal meshes are not considered subsequently. In consequence, the terms mesh and conformal mesh are used synonymously throughout the rest of the book.

A mesh $M := (P, E)$ is defined by the set $P := \{p_i \in \mathbb{R}^m : i = 1, \ldots, n_p\}$ of n_p nodes p_i and the set $E := \{E_i^{\text{type}} : i = 1, \ldots, n_e\}$ of n_e elements E_i^{type}. These elements do not necessarily have to be of the same type. Usually, in a mesh context, the element is defined by a node index tuple, i.e., $E_i^{\text{type}} := (j_{i,1}, \ldots, j_{i,n_{\text{type}}})$, with pairwise distinct node indices $j_{i,k} \in \{1, \ldots, n_p\}$ and

$$P = \begin{pmatrix} 0 & 0 \\ 2 & 0 \\ 2 & 2 \\ 0 & 2 \\ 1 & 1 \end{pmatrix}, \quad E = \begin{pmatrix} 1 & 2 & 5 \\ 2 & 3 & 5 \\ 3 & 4 & 5 \\ 4 & 1 & 5 \end{pmatrix}$$

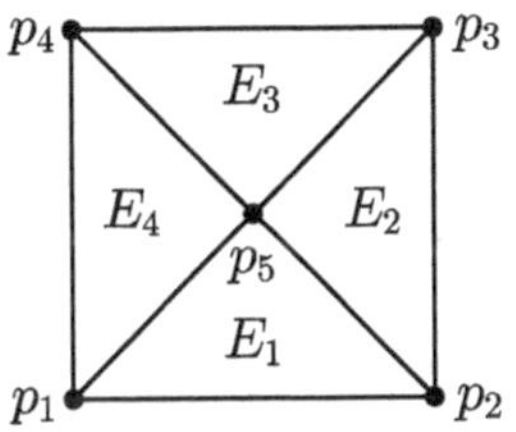

Figure 2.3: Matrix representation of a triangular mesh $M = (P, E)$ with nodes p_i and elements E_i given by the rows of P and E, respectively.

n_{type} being element specific. Here, it is assumed that all elements of a specific type share the same node numbering scheme. Such schemes have been given in the previous section for the most common two- and three-dimensional elements.

It should be noticed that this element definition based on node index tuples differs from the element representation based on node matrices, which has been used in Section 2.1.2. Both representations will be used synonymously throughout the rest of this book with the one exception that if elements occur in equations, their node matrix representation is used.

Meshes with only one element type can conveniently be defined alternatively using matrix representations for P and E. Let $p_{i,k}$ denote the kth coordinate of the node p_i. Then the $n_p \times m$ node matrix is given by $P := (p_{i,k})$ and the $n_e \times n_{\text{type}}$ element node index matrix is given by $E := (j_{i,k})$.

Figure 2.3 shows an example of a planar triangular mesh with $n_p = 5$ nodes and $n_e = 4$ elements. The ith row of P contains the coordinates of the ith node. Similarly, the ith row of E provides the node indices of the ith element.

2.2.2 Mesh types

Meshes can be distinguished by their topological structure and distribution of elements over the given domain Ω. For example, *structured meshes* are characterized by a regular connectivity of nodes and elements. They can, for example, be generated by the transformation of a regular grid as shown in Figure 2.4a. In this, each of the inner nodes is connected to four elements, and each of the outer nodes, also denoted as boundary nodes, is connected to two elements. Furthermore, each inner element has four neighbors sharing a common edge and four neighbors only sharing one common node. Each outer element has three edge-connected neighbors and two node-connected neighbors. Since the mesh is a projection of a regular grid, it can be stored very efficiently.

In contrast, in *unstructured meshes* the connectivity of nodes and elements varies, which provides more flexibility in meshing complex domains, while preserving element size. Such a mesh is depicted in Figure 2.4b. It consists of 359 triangular elements and 223 nodes. The number of elements connected to one node ranges from two to eight.

The *mesh density* describes the average number of elements per unit area or

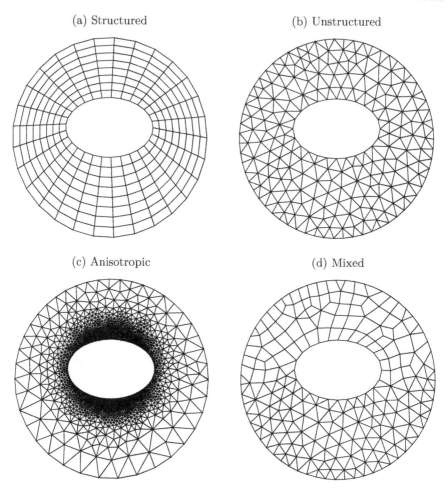

Figure 2.4: Types of conformal meshes.

volume. Meshes with a uniform density, as depicted in Figure 2.4b, are called *isotropic* meshes. In contrast, if the mesh density varies in a given direction, the mesh is called *anisotropic*. Such a mesh is depicted in Figure 2.4c. As can be seen, mesh density increases towards the inner boundary of the domain. Usually, anisotropic meshes are used in a context where the geometry or the solution to be approximated over the given domain changes rapidly. This is taken into consideration by a local refinement of regions, where such features have to be modeled.

If a mesh consists of elements of different types, it is called a *mixed mesh*. For example, Figure 2.4d depicts a mesh consisting of 199 triangular and 80 quadrilateral elements, which have been generated from the mesh depicted in Figure 2.4b by merging 80 pairs of triangles sharing a common edge. Here some triangular elements are enclosed by quadrilateral elements in the upper half of the mesh, whereas the lower part of the mesh consists entirely of triangular elements.

If the upper half of the mesh depicted in Figure 2.4d consisted entirely of quadrilateral elements, the resulting mesh would consist of a triangular and a quadrilateral submesh glued together. Such mixed meshes, consisting of single element type submeshes sharing common interfaces, are also called *hybrid meshes*. Examples of such meshes are considered in Section 7.3.1.

Finally, a mesh is called *simplicial* if it consists entirely of simplicial elements of the same dimension. Simplicial elements are, for example, triangles in 2D and tetrahedra in 3D.

2.3 Quality criteria

2.3.1 Introduction

Mesh validity and quality play a vital role in finite element analysis and the underlying approximation theory, since they have a direct influence on solvability and solution accuracy. A broad variety of element quality criteria have been proposed in the literature, ranging from geometric measures and relations to more elaborate algebraic approaches. The *quality criterion* maps an arbitrary mesh element E to its *quality number* $q(E) \in \mathbb{R}$. As a minimal requirement, it must be invariant under rotation, scaling, and translation. Furthermore, it should be general in the sense that it is applicable to all types of elements under consideration. From a computational point of view, it is favorable if $q(E)$ is normalized to $[0, 1]$, where the quality number zero indicates degenerate elements and the quality number one indicates regular elements. Furthermore, quality criteria can be distinguished by other criteria, such as numerical complexity of evaluation, or algebraic properties like differentiability, etc.

For example, the *edge length ratio*, defined as the length of the shortest edge of the element divided by the length of the longest edge of the element, does not represent a proper quality criterion. Although the ratio is applicable to all types of elements, its values are within $[0, 1]$, and it is invariant under all transformations mentioned before, it does not reach its maximum only for the regular element. This can be seen in the example of a planar quadrilateral element. Here, not only the square yields edge length ratio one, but also each parallelogram consisting of edges of the same length. This even holds for the degenerate parallelogram with two interior angles being zero and the others being π. Hence, the edge length ratio is not a suitable quality criterion for judging element regularity.

2.3.2   The mean ratio quality criterion

Throughout the rest of this book, the well-known *mean ratio quality* criterion is used. It has been developed in the context of optimization-based smoothing methods [53,84], and is based on measuring the distance of an arbitrary simplex in $\mathbb{R}^m$ from a corresponding reference simplex, where the reference simplex depends on the element type.

Let $E := (p_1, \ldots, p_{|E|})$ denote an element with $|E|$ nodes, which can be either triangular or quadrilateral in $\mathbb{R}^2$, or tetrahedral, hexahedral, pyramidal, or prismatic in $\mathbb{R}^3$. Except for the apex of the pyramidal element, each node $p_i \in \mathbb{R}^m$, with $m \in \{2, 3\}$, of any element, is edge-connected to m other nodes p_j of the same element. Thus, each node and its edge-connected neighbors define a simplex in $\mathbb{R}^m$, which will be called a *node simplex*.

Figure 2.5 gives examples of such simplices associated to p_1, which are marked gray for all element types under consideration. Here, the given elements are regular, and the first two nodes have been set to $p_1 = (0,0)^t$ and $p_2 = (1,0)^t$, or $p_1 = (0,0,0)^t$ and $p_2 = (1,0,0)^t$ in the case of $m = 2$ or $m = 3$, respectively. As a prerequisite of the definition of the mean ratio quality criterion, specific element validity conditions have to be stated first. Element validity will also be assessed based on the node simplices, which will be rendered more precisely first. In the following, an element of a specific type will be denoted as E^{tri}, E^{quad}, E^{tet}, E^{hex}, E^{pyr}, or E^{pri}, respectively. The node indices of the node simplices for the corresponding element type are given by the tuples

$$N^{\mathrm{tri}} := \big((1,2,3)\big),$$
$$N^{\mathrm{quad}} := \big((1,2,4),(2,3,1),(3,4,2),(4,1,3)\big),$$
$$N^{\mathrm{tet}} := \big((1,2,3,4)\big),$$
$$N^{\mathrm{hex}} := \big((1,4,5,2),(2,1,6,3),(3,2,7,4),(4,3,8,1),(5,8,6,1),(6,5,7,2),$$
$$(7,6,8,3),(8,7,5,4)\big),$$
$$N^{\mathrm{pyr}} := \big((1,2,4,5),(2,3,1,5),(3,4,2,5),(4,1,3,5)\big),$$
$$N^{\mathrm{pri}} := \big((1,2,3,4),(2,3,1,5),(3,1,2,6),(4,6,5,1),(5,4,6,2),(6,5,4,3)\big).$$

In this representation, the kth element of $N \in \{N^{\mathrm{tri}}, N^{\mathrm{quad}}, N^{\mathrm{tet}}, N^{\mathrm{hex}}, N^{\mathrm{pyr}}, N^{\mathrm{pri}}\}$ represents the $m + 1$ node indices $N_{k,i}$, $i \in \{1, \ldots, m + 1\}$ of the node simplex associated with p_k. The number of node simplices, i.e., the number of elements in the tuple N, is denoted as $|N|$. It should be noticed that in the case of E^{tri} and E^{tet} each of the $m + 1$ node simplices represents the element itself. Hence it suffices to use only the node simplex associated with p_1, resulting in $|N^{\mathrm{tri}}| = |N^{\mathrm{tet}}| = 1$. Furthermore, the apex p_5 of the pyramidal element is the only node with an incident edge number that is not equal to m. Therefore, this node is omitted, since only the four tetrahedra associated to the base nodes will be used in order to define the mean ratio quality number of a pyramidal element. Consequently, for the triangular and tetrahedral as well as the pyramidal element, it holds that $|E| \neq |N|$.

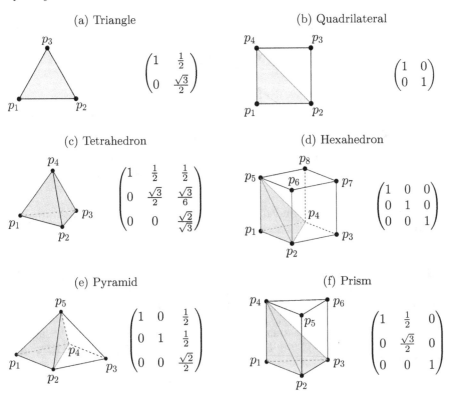

Figure 2.5: Regular elements with reference simplices marked gray and associated weight matrices W.

Definition 2.1 (Element validity). *Let E denote an arbitrary element taken from $\{E^{\mathrm{tri}}, E^{\mathrm{quad}}, E^{\mathrm{tet}}, E^{\mathrm{hex}}, E^{\mathrm{pyr}}, E^{\mathrm{pri}}\}$ with its associated node simplices indices tuple given by $N \in \{N^{\mathrm{tri}}, N^{\mathrm{quad}}, N^{\mathrm{tet}}, N^{\mathrm{hex}}, N^{\mathrm{pyr}}, N^{\mathrm{pri}}\}$. For each node p_k, $k \in \{1, \dots, |N|\}$, let*

$$T_k := \left(p_{N_{k,1}}, \dots, p_{N_{k,m+1}}\right) \tag{2.5}$$

be the associated node simplex, *and*

$$D(T_k) := \left(p_{N_{k,2}} - p_{N_{k,1}}, \dots, p_{N_{k,m+1}} - p_{N_{k,1}}\right)$$

the $m \times m$ matrix of its spanning edge vectors. The element is called valid *if $\det(D(T_k)) > 0$ holds for all $k \in \{1, \dots, |N|\}$. Otherwise it is called* invalid.

The given validity criterion excludes specific cases of degenerated elements, like, for example, tetrahedra with collinear nodes. This definition of element validity is also a prerequisite of the finite element requiring a positive Jacobian.

In the following, a definition of the mean ratio quality criterion is given,

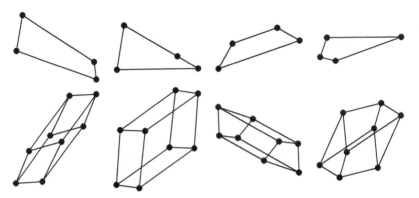

Figure 2.6: Quadrilateral and hexahedral elements with mean ratio quality number $q(E) \approx 1/2$.

which is based on measuring the deviation of simplices in an arbitrary valid element from their counterpart in an ideal, i.e., regular element, which is represented by a target matrix W as defined in [84].

Definition 2.2 (Element quality). *Let E denote an arbitrary valid element taken from $\{E^{\text{tri}}, E^{\text{quad}}, E^{\text{tet}}, E^{\text{hex}}, E^{\text{pyr}}, E^{\text{pri}}\}$, with nodes in $\mathbb{R}^m$, and $N \in \{N^{\text{tri}}, N^{\text{quad}}, N^{\text{tet}}, N^{\text{hex}}, N^{\text{pyr}}, N^{\text{pri}}\}$ its associated indices tuple of node tetrahedra T_k, according to (2.5). The* mean ratio quality number *of the element is given by*

$$q(E) := \frac{1}{|N|} \sum_{k=1}^{|N|} \frac{m \det(S_k)^{2/m}}{\text{trace}(S_k^{\text{t}} S_k)}, \quad S_k := D(T_k) W^{-1}, \tag{2.6}$$

with the element type-dependent target simplex weight matrix W given in Figure 2.5. Here, the columns of W represent the spanning edge vectors of a node tetrahedron in the associated regular element.

It holds that $q(E) \in [0, 1]$, with small quality numbers representing low quality elements, and $q(E) = 1$ representing ideal regular elements. In the case of pyramidal elements only the node tetrahedra associated to the four base nodes are used for quality computation, which is in accordance with [99]. Furthermore, in the case of non-simplicial elements, the associated difference matrices W represent non-regular simplices. It should also be noticed that the denominator in (2.6) represents the squared Frobenius norm of the matrix S_k. For simplicity reasons, the quality number of invalid elements is also set to zero in the following.

Each node coordinate of an element is a degree of freedom. Hence, representing a single element by a real-valued quality number, it is obvious that different elements sharing the same quality number can differ in shape. This is illustrated in Figure 2.6, which depicts random node elements with a quality number deviating less than 10^{-3} from $1/2$. As can be seen, the shapes of the elements of the same type differ significantly. However, the quality number $q(E) = 1$ is only achieved for the regular element in all cases.

2.3.3 Mesh quality numbers

Based on the mean ratio quality number for elements, the following mesh quality numbers are defined.

Definition 2.3 (Mesh quality). *Let M denote a mesh with n_e valid elements E_i, $i \in \{1, \ldots, n_e\}$. The min mesh quality number q_{min} representing the lowest quality number of all mesh elements of M and the mean mesh quality number q_{mean} representing the arithmetic mean of all mesh element quality numbers of M are given by*

$$q_{min} := \min_{i \in \{1, \ldots, n_e\}} q(E_i) \quad and \quad q_{mean} := \frac{1}{n_e} \sum_{i=1}^{n_e} q(E_i), \tag{2.7}$$

respectively.

The definitions given by (2.7) imply $q_{min} \leq q_{mean}$. Furthermore, due to $q(E_i) \in [0, 1]$, it follows directly that $q_{min} \in [0, 1]$ and $q_{mean} \in [0, 1]$. In both cases, larger quality numbers indicate better mesh quality. In particular, $q_{min} = 1$ implies $q_{mean} = 1$, representing a mesh consisting only of regular elements. Examples of such meshes are Platonic meshes. The mesh quality numbers of the meshes depicted in Figure 2.4 are given in Table 2.2.

Table 2.2: Quality numbers of the meshes depicted in Figure 2.4.

Mesh	q_{min}	q_{mean}
Structured	0.5796	0.8112
Unstructured	0.7887	0.9451
Anisotropic	0.0196	0.8023
Mixed	0.6224	0.9404

It should be noticed that the definition of mesh regularity is usually not based on element regularity. A mesh element is called regular if its mean ratio quality number is one indicating the highest level of symmetry. However, meshes are also called regular if all inner nodes share the same topological configuration with respect to the adjacent elements. That is, the terms structured mesh and regular mesh are used synonymously.

2.4   Mesh generation

2.4.1   Overview

One step of the finite element method consists of tessellating the simulation domain into simple finite elements, like triangles and quadrilaterals in 2D or tetrahedra and hexahedra in 3D. These finite elements are then used to define the supports of trial functions approximating the solution of the boundary value problem to be solved. It is well known that the finite element solution accuracy and computational efficiency depend on the size and shape of the elements, and hence on the quality of the underlying mesh [131].

For example, large element angles lead to large errors in the gradient, whereas small angles can significantly increase the condition of the stiffness matrix [6, 125], impairing the solution accuracy and the convergence properties of iterative solvers. Beyond that, requirements for finite element meshes depend heavily on the type of application [12]. Therefore, in the finite element simulation process, mesh generation techniques play a vital role and hence are a main topic of research. In the following, a brief overview of mesh generation techniques will be given based on [55,91,93,105]. Readers who are interested in a profound and comprehensive introduction to mesh generation are referred to the excellent books of Lo [93] or Frey and George [55].

Structured meshes are characterized by the property that all interior nodes of the mesh have an equal number of adjacent elements. A basic example of such meshes is a Cartesian grid consisting of quadrilateral or hexahedral elements only. The main idea of all *structured mesh generation* techniques is to generate a grid for canonical domains and to map the resulting mesh to a physical domain defined by its boundary discretization [55]. This was also the approach in generating the mesh depicted in Figure 2.4a. The generation of structured meshes becomes difficult if the complexity of the physical domain increases. Here, *domain decomposition* techniques can be applied to partition the complex domain into a set of simple subdomains in order to reduce the problem to local mappings. However, the automation of such processes is not straightforward, and meshing constraints imposed by the shared interface meshes of subdomains must be considered.

In contrast, unstructured meshes allow any number of elements to meet a single node and therefore provide a significantly greater flexibility in meshing arbitrarily complex domains. Therefore, these meshes are of greater practical importance, and the following will address common unstructured mesh generation techniques only.

For simplicial mesh generation, that is, meshes consisting entirely of simplices, the *quadtree or octree method* [121, 171, 172] uses hierarchical tree structures obtained by recursive subdivision of a square or cube containing the domain or surface until the desired resolution is reached. On each level, the obtained cells are classified into outer cells, which can be omitted, irregular cells, where boundary intersections must be computed, and inner cells. On the last recursion level, irregular and inner cells are then meshed into triangles or tetrahedra. To ensure that element sizes do not change too dramatically, the maximum difference in subdivision levels between adjacent cells can be limited, hence leading to balanced trees. In addition, subsequent mesh improvement steps are accomplished to increase mesh quality [11, 54].

In contrast, *advancing front* techniques progressively build elements inward starting from the boundary. This iterative technique proceeds by placing new nodes and the generation of new elements, which are connected to the existing elements, until the entire domain is meshed [75, 94, 96]. Common steps in advancing front techniques are: front analysis, internal point creation, element generation, validation, convergence control, and mesh improvement [55]. Due to the high-quality point distribution, the advancing front method is used in many commercial meshing software packages [105].

Mesh generation by *Delaunay triangulation* belongs to the most popular simplicial mesh generation techniques, due to their simplicity, reliability, efficiency, and other favorable properties [36, 55, 105, 123]. Common steps in Delaunay-based mesh generation are: initial triangulation of the bounding box of the domain, boundary points insertion into the triangulation, original boundary recovery, internal point creation or insertion, and finally, mesh improvement. The triangulation process is based on the *empty sphere criterion* [30] stating that any node must not be contained within the circumsphere of any simplex within the mesh.

One of the remarkable properties of the Delaunay triangulation is its uniqueness for a *non-degenerated* set of nodes. For planar triangulations, a set of nodes is called non-degenerated if no three nodes are on the same line and no four nodes are on the same circle [124]. An essential property from a finite element application point of view is the fact that the minimum angle in a Delaunay triangulation is greater than the minimum angle in any other triangulation of the same nodes [124]. Computing the Delaunay triangulation of a set of n_p nodes can be accomplished in $\mathcal{O}(n_p \log n_p)$, as has been shown in [88]. A modified divide-and-conquer approach using pre-partitioning was shown in [37] to run in $\mathcal{O}(n_p \log \log n_p)$ expected time while retaining the optimal $\mathcal{O}(n_p \log n_p)$ worst-case complexity. Alternative methods exist, which tetrahedralize volumetric models by a recursive subdivision approach [31, 111].

Hexahedral meshes are preferred over tetrahedral meshes in specific finite element and finite volume applications, like, for example, elastic or elastic-plastic solid continuum analysis or CFD applications, since more suitable trial functions can be defined on such tessellations, resulting in a higher solution accuracy combined with a lower number of mesh elements [10, 24]. However, depending on the geometrical complexity of the simulation domain, *all-hexahedral mesh generation* is in general considerably harder than tetrahedral mesh generation. Hence a variety of hexahedral mesh generators, as described, for example, in [89, 109, 133], are built on subdividing the initial domain into primitives. The latter can subsequently be meshed using, for example, *sweeping*, i.e., by moving a quadrilateral mesh from a source surface along a given path to a target surface while generating hexahedral elements [82,114,122]. Alternative subdomain meshing techniques can be based on *octrees* [76,121,174], advancing-front-like approaches, such as *plastering* [129] or *Whisker weaving* [47,134]. The latter is based on the concept of the spatial twist continuum, which is the dual of the hexahedral mesh represented by an arrangement of intersecting surfaces that bisect hexahedral elements in each direction [105]. Furthermore, various specialized algorithms exist. For example, an approach to mesh composite domains consisting of heterogeneous materials with non-manifold region surfaces is described in [176].

Since meshing complex volume models by hexahedra is considerably harder than meshing by tetrahedra, the achieved degree of generality and automation is still far from the state obtained in the case of tetrahedral meshing. Therefore, the use of *hybrid meshes* and corresponding mesh generators has been proposed in order to combine the advantages of tetrahedral and hexahedral elements [46,74,165]. In addition, pyramids and prisms serve as transition elements to connect triangular and quadrilateral element faces. Depending on the application, hybrid meshes combining other types of elements might be preferred; these include tetrahedral meshes with prismatic boundary layers in computational biofluid dynamic computations [38].

2.4.2   Delaunay meshes

Delaunay triangulations are simplicial meshes presented in [30]. They play a prominent role in mesh generation, since they are defined for arbitrary dimensions, efficient and simple algorithms exist, and they build a good foundation for quality mesh generation. Due to the importance of this mesh generation technique for finite element applications, the following gives a more detailed description of Delaunay meshing and the properties of Delaunay triangulations.

For a set P of n_p nodes $p_i \in \mathbb{R}^m$, $i \in \{1, \ldots, n_p\}$, the Delaunay triangulation is a tessellation of the convex hull of P into simplicial elements E_j, each consisting of $m+1$ nodes fulfilling the *empty sphere criterion*. This criterion states that no point of P is inside the circum-hyper-sphere of any simplex E_j. Furthermore, the Delaunay triangulation of P is unique, if no subset of $m+2$ nodes of P lies on the boundary of a sphere not containing any other node of P inside it. Nodes fulfilling this criterion are denoted as being in *general position*.

Figure 2.7a depicts ten nodes in general position and the associated unique Delaunay triangulation. The empty sphere criterion is illustrated with the example of an inner triangle. As can be seen, the circumcircle of the triangle with nodes marked by circles does not contain any other node of the triangulation marked by discs. Since this holds for the circumcircle of all triangles of the mesh, the given triangulation is the Delaunay triangulation. In contrast, the triangulation of the same set of nodes depicted in Figure 2.7b does not fulfill the

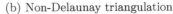

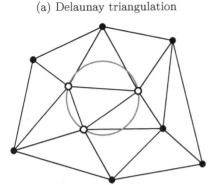
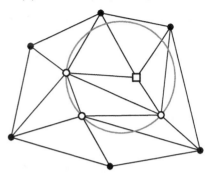

Figure 2.7: The empty sphere criterion.

empty sphere criterion. This is also shown in the example of the circumcircle of an inner triangle with nodes marked by circles. This circumcircle marked gray does contain a mesh node marked by a square and hence does not fulfill the empty sphere criterion.

By comparing the two meshes, it can be observed that the Delaunay triangulation depicted on the left does consist of better quality triangles. In particular, the minimal element angle is larger, if compared to the minimal element angle of the non-Delaunay mesh depicted on the right. One of the remarkable properties of the Delaunay triangulation is that it maximizes the minimal element angle for all triangulations of the node set P. Hence Delaunay meshing results in reasonable quality meshes. However, the maximum angle is not necessarily minimized [40].

A broad variety of Delaunay triangulation methods have been proposed, being, for example, based on *flipping the common edge* of two neighboring triangles in order to fulfill the empty sphere criterion, iterative *node insertion* like the algorithms of Bowyer and Watson [13, 166], or *divide and conquer* schemes. Here, planar Delaunay triangulation based on edge flipping is of complexity $\mathcal{O}(n_p^2)$. In contrast, divide and conquer algorithms usually are of complexity $\mathcal{O}(n_p \log n_p)$, allowing a very efficient computation of Delaunay triangulations. An almost linear complexity can be achieved for certain types of node sets, such as evenly distributed randomly generated nodes [93]. Delaunay triangulations can be obtained as the geometric dual of *Voronoi tessellations*. That is, the Voronoi diagram of P is computed and two nodes of P share a common edge in the Delaunay triangulation, if the associated Voronoi cells share a common edge [93, 167].

Delaunay triangulation algorithms can readily be used for finite element mesh generation, as is depicted in Figure 2.8. One approach is to determine a discretization of the boundary of the simulation domain Ω to be meshed first. This is depicted in Figure 2.8a. Here, the nodes marked by black discs are placed almost equidistantly along the inner and outer curves of the boundary $\partial\Omega$

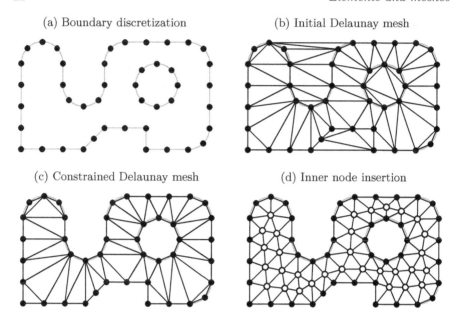

(a) Boundary discretization   (b) Initial Delaunay mesh

(c) Constrained Delaunay mesh   (d) Inner node insertion

Figure 2.8: Delaunay meshing of a planar domain.

marked gray. Delaunay triangulation of this point set results in the triangular mesh depicted in Figure 2.8b. As can be seen, the union of all triangular elements of this Delaunay triangulation of the node set P results in the convex hull of P. Since Ω is not convex, triangles have been constructed that are located outside of the polygonal approximation of $\partial\Omega$. By removing these outside triangles from the mesh, the so-called *constrained Delaunay triangulation* is derived. Depending on the discretization width of $\partial\Omega$, some edges of the polygonal approximation of $\partial\Omega$ may not be contained in this Delaunay triangulation. In this case, the mesh must be locally modified, which is known as *boundary recovery*.

In the given example, the constrained Delaunay triangulation has been obtained by removing triangles with centroids not being in Ω, resulting in the mesh depicted in Figure 2.8c. As can be seen, this mesh is not suitable for finite element approximation, since it consists of boundary nodes only and since it is of comparably low quality. Nevertheless, *point insertion* algorithms can readily be used to result in quality finite element meshes like the one depicted in Figure 2.8d. This mesh has been generated by iteratively inserting inner nodes in the mesh, while preserving the empty sphere property. Here, the iterative placement of inner nodes marked by circles is usually based on local element size and quality. To improve the quality of the resulting mesh, smoothing techniques are usually applied after mesh generation. More detailed descriptions of Delaunay triangulation-based mesh generation approaches can be found in [93].

3

The finite element method

A brief history of the finite element method is given first in this chapter. After that, fundamentals are examined with the example of an elliptical model problem. To do so, Sobolev spaces, the variational formulation, and the Ritz-Galerkin method are introduced, which leads to an approximate finite element solution of the model problem. Furthermore, the influence of mesh quality on solution accuracy and stability is discussed through the example of an iteratively distorted mesh.

3.1 History of the finite element method

The *finite element method (FEM)* is a versatile numerical technique for computing approximate solutions of boundary value problems for partial differential equations. Categories of application are equilibrium problems, eigenvalue problems, and time-dependent problems originating in the context of structural analysis, heat transfer, fluid mechanics, electromagnetism, biomechanics, geomechanics, and acoustics. Key features of the finite element method are [180]: (a) the continuum is divided into a finite number of parts (elements), the behavior of which is specified by a finite number of parameters, and (b) the solution of the complete system as an assembly of its elements. Developing the finite element method as it is known today was a long evolutionary process in which various engineers, physicists, and mathematicians provided major contributions over several decades [25, 64, 73, 116].

During World War II, Argyris invented the finite element technique in the course of the stress analysis of the swept-back wing of the twin engine Meteor Jet Fighter [5, 135]. In 1954 and 1955 he published a series of articles, also republished in the monograph [4], in which the matrix theory of structures was developed for discrete elements. It was shown that this is only a special case of the general continuum in which stresses and strains have been specified, which leads to the concept of flexibility and stiffness. Argyris recognized the concept of duality between equilibrium and compatibility and provided equations relating stresses and strains to loads and displacements [64]. In the historical overview [25], Clough stated in 2004: "In my opinion, this monograph (...)

certainly is the most important work ever written on the theory of structural analysis, ..."

Independent of the work of Argyris and from a mathematical point of view, Courant developed in [27] the idea of minimizing a functional using linear approximation over subregions, with the values being specified at discrete points, which in essence became the node points of a mesh of elements. Courant solved the Saint-Venant's torsion of a square hollow box and in the appendix of his paper he introduced the idea of linear approximation over triangular areas [64]. Courant's contribution is based on earlier results by Rayleigh [110] and Ritz [112].

Under the guidance of Turner, in 1952 Clough started to work on in-plane stiffness matrices for 2D plates for analyzing the vibration properties of delta wings for Boeing. This resulted in solving plane stress problems by means of triangular elements whose properties were determined from the equations of elasticity theory as reported in [137]. This publication also addresses the question of convergence in the case of mesh refinement [64] and introduced the direct stiffness method for determining finite element properties [73]. Results of further treatments of Clough on the plane elasticity problem are reported in the publication [26], which also gives the verification that for known geometries and loading the stresses converged to the corresponding analytic solution [64]. This publication of Clough also introduced the name finite element method. As an explanation for his choice, Clough stated in a speech on the early history of the FEM given in [25]: "On the basis that a deflection analysis done with these new 'pieces' (or elements) of the structure is equivalent to the formal integration procedure of integral calculus, I decided to call the procedure the FEM because it deals with finite components rather than differential slices."

The first book on finite elements was published in 1967 by Zienkiewicz. The revised version [180] of this book is still one of the standard references for the finite element method. Being mainly introduced in the context of structural analysis, Zienkiewicz clarified the connection to function minimization techniques and opened the way to the analysis of field problems by the FEM [64]. Revealing the broad applicability of the finite element method, this was also the starting point for vital research by the scientific community not only with respect to fields of application but also with respect to the efficiency, effectivity, and implementational aspects of this method.

This is indicated by Figure 3.1, which depicts the total number of scientific documents published so far over the last decades containing the phrase "finite element" in their title. Results are given for the Google Scholar database [61] as of November 5, 2017. This specific search does not cover the additional multitude of FEM publications omitting these key words in their title. Nevertheless, these numbers impressively show the vivid development of the finite element method and its applications for more than 50 years.

In recent decades, highly effective mesh-less and mesh-free methods have been developed. They eliminate the need for meshing the simulation domain by using domain independent regular background grids or scattering approaches.

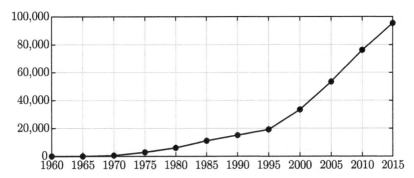

Figure 3.1: Total number of publications containing the phrase "finite element" in their title.

Such methods can, for example, be based on WEB-splines or radial basis functions [7, 67–72, 92, 102]. However, mesh-based methods still have their specific advantages and are therefore common practice in many fields of application. They are implemented by a broad variety of robust, efficient, and flexible commercial as well as open source computer programs, which generate a high demand for quality mesh generation and mesh improvement.

3.2 Fundamentals

This section describes the fundamentals of the finite element method on the example of elliptical boundary value problems of second order. The presentation given here is based on that given in the books [14, 15, 168]. To tighten things up, most of the proofs have been omitted. Readers interested in a more in-depth overview of the mathematical background of the finite element method are referred to the literature cited above. Readers primarily interested in the computational aspects are referred to Sections 3.2.1 and 3.2.4.

3.2.1 Elliptical boundary value problems

Let Ω denote a *domain*, i.e., an open connected and bounded subset of $\mathbb{R}^m$ with $m \in \mathbb{N}$. Its *boundary* $\partial\Omega := \overline{\Omega}\backslash\Omega$ is obtained by subtracting Ω from its closure $\overline{\Omega}$. For a given function $f : \Omega \to \mathbb{R}$, the *general partial differential equation of second order*, with constant coefficients $a_{k,\ell}, b_k, c \in \mathbb{R}$ is given by

$$-\sum_{k=1}^{m}\sum_{\ell=1}^{m} a_{k,\ell} \frac{\partial^2 u}{\partial x_k \partial x_\ell} + \sum_{k=1}^{m} b_k \frac{\partial u}{\partial x_k} + cu = f. \tag{3.1}$$

For a two-times continuously differentiable solution function $u : \overline{\Omega} \to \mathbb{R}$, the matrix $A := (a_{k,\ell})$ is symmetric due to the theorem of Schwarz, stating that the partial derivatives are commutative. If the matrix A is *positive definite*, i.e., $x^t A x > 0$ for all $x \in \mathbb{R}^m \backslash \{0\}$, then (3.1) is called an *elliptic partial differential equation of second order*. For an accordingly defined differential operator L, this partial differential equation can be written as $Lu = f$.

The existence and uniqueness of a solution of (3.1) depends on the proper choice of additional boundary conditions. For elliptic partial differential equations, the following two admissible basic types are of special interest:

- *Dirichlet boundary condition*: Prescribed values on the boundary, i.e., $u = g$ for all $x \in \partial\Omega$.

- *Neumann boundary condition*: Prescribed derivative along the boundary, i.e., $\partial u / \partial n = g$ for all $x \in \partial\Omega$, where n denotes the exterior normal to the boundary $\partial\Omega$.

Let $\delta_{k,\ell}$ denote the *Kronecker symbol* with $\delta_{k,\ell} := 1$ for $k = \ell$ and $\delta_{k,\ell} := 0$ for $k \neq \ell$. The canonical choice $a_{k,\ell} = \delta_{k,\ell}$ and $b_k = c = 0$ results in the *Laplace operator* $L = -\Delta := -\operatorname{div}\operatorname{grad}$. Together with Dirichlet boundary conditions, this leads to the following prototype of an elliptic boundary value problem of second order.

Definition 3.1 (Model problem). *For a domain $\Omega \subset \mathbb{R}^m$ and the right-hand side $f : \Omega \to \mathbb{R}$, the* Poisson problem *with homogeneous Dirichlet boundary conditions is given by*

$$
\begin{aligned}
-\Delta u &= f & \text{in } \Omega, \\
u &= 0 & \text{on } \partial\Omega.
\end{aligned}
\tag{3.2}
$$

A Poisson problem with *inhomogeneous Dirichlet boundary condition $u = g$* on $\partial\Omega$ can be transformed to a Poisson problem with homogeneous Dirichlet boundary conditions, i.e., zero boundary conditions, if there exists a suitable function $\widehat{g} : \overline{\Omega} \to \mathbb{R}$ that equals g if restricted to the boundary $\partial\Omega$. For such a function, the transformation $w := u - \widehat{g}$ leads to the alternative Poisson problem $-\Delta w = f + \Delta \widehat{g}$ with homogeneous Dirichlet boundary condition $w = 0$ on $\partial\Omega$.

The solution $u(x)$ of the Poisson problem with homogeneous Dirichlet boundary problem can be interpreted as the pointwise displacement of a membrane at $x \in \Omega$ under the transversal force $f(x)$. Here, the homogeneous Dirichlet boundary condition represents the clamping of the membrane along $\partial\Omega$ [17]. For the constant load $f \equiv 1$, this is shown exemplarily on the right of Figure 3.2 for the planar domain Ω depicted on the left.

A solution $u \in C^2(\Omega) \cap C^0(\overline{\Omega})$ of (3.2), i.e., a solution that is two times continuously differentiable in Ω and continuous on its closure $\overline{\Omega}$, is denoted as a *classical solution*. For the existence of such a solution, additional constraints on f, g, and the shape of $\partial\Omega$ are required. In addition, such classical solutions are generally hard to find analytically [60]. Hence, solutions are sought in more general function spaces, which will be introduced in the next section.

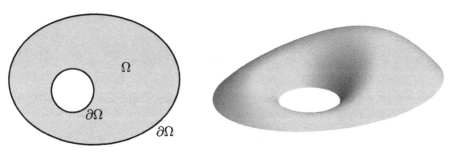

Figure 3.2: Domain Ω and side view of the displaced membrane.

3.2.2 Sobolev spaces

The key to defining more general function spaces lies in the generalization of the Riemann integral by the Lebesgue integral. The latter will serve as a selection criterion for a function space according to the following definition.

Definition 3.2 (Lebesgue space). *The function space*

$$L^2(\Omega) := \left\{ u : \Omega \to \mathbb{R} : \int_\Omega u^2(x)\, dx < \infty \right\}$$

of all quadratic Lebesgue integrable functions u defined on a domain $\Omega \subset \mathbb{R}^m$ is called Lebesgue space. *A scalar product and its associated norm are given on $L^2(\Omega)$ by*

$$\langle u, v \rangle_{L^2(\Omega)} := \int_\Omega u(x)v(x)\, dx \quad and \quad \|u\|_{L^2(\Omega)} := \sqrt{\langle u, u \rangle_{L^2(\Omega)}}. \qquad (3.3)$$

The function space $L^2(\Omega)$ is complete with respect to the norm $\|\cdot\|_{L^2(\Omega)}$ and forms in combination with the scalar product $\langle \cdot, \cdot \rangle_{L^2(\Omega)}$ a Hilbert space [18]. However, this space is still not suitable for the boundary value problem (3.2), since additional assumptions on the differentiability of its elements must be imposed.

In $L^2(\Omega)$ two functions $u, v : \Omega \to \mathbb{R}$ are identified, if their function values only differ on a set of measure zero, i.e., it holds that $\|u - v\|_{L^2(\Omega)} = 0$. Similarly, a more general definition of differentiation is sought. It will be introduced using both the classic definition of differentiation and the Lebesgue integral. Therefore, let

$$\operatorname{supp} v := \overline{\{x \in \mathbb{R}^m : v(x) \neq 0\}}$$

denote the *support* of the function v and $C_0^\infty(\Omega)$ the set of functions differentiable for all degrees of differentiation, also called smooth functions, with compact support in Ω. For the *multi-index*

$$\alpha := (\alpha_1, \ldots, \alpha_m) \in \mathbb{N}_0^m$$

the *classic differential operator* is defined as

$$\widehat{D}^\alpha := \frac{\partial^{|\alpha|}}{\partial x_1^{\alpha_1} \cdots \partial x_m^{\alpha_m}} \quad \text{with order} \quad |\alpha| := \sum_{\nu=1}^{m} \alpha_\nu.$$

With the aid of the Lebesgue integral, the following definition will provide the generalization of this differential operator.

Definition 3.3 (Weak derivative)**.** *Let Ω be a domain in $\mathbb{R}^m$, $\alpha \in \mathbb{N}_0^m$ and $u \in L^2(\Omega)$. If there exists a function $v \in L^2(\Omega)$ with*

$$\langle w, v \rangle_{L^2(\Omega)} = (-1)^{|\alpha|} \left\langle \widehat{D}^\alpha w, u \right\rangle_{L^2(\Omega)} \quad \forall w \in C_0^\infty(\Omega),$$

then $D^\alpha u := v$ will be denoted as a weak derivative *of u.*

It can be shown by partial integration that for $u \in C^{|\alpha|}(\Omega)$ the classic derivative $\widehat{D}^\alpha u$ equals the weak derivative $D^\alpha u$ [18]. The function spaces required for solving the model problem (3.2) will now be introduced by embedding the generalized differentiation and accordingly modified norms in L^2.

Definition 3.4 (Sobolev spaces H^ℓ)**.** *Let $\Omega \subset \mathbb{R}^m$ be a domain and $\ell \in \mathbb{N}_0$. On the* Sobolev space

$$H^\ell(\Omega) := \left\{ u \in L^2(\Omega) \ : \ D^\alpha u \text{ exists } \forall \alpha \in \mathbb{N}_0^m \text{ with } |\alpha| \leq \ell \right\} \tag{3.4}$$

of order ℓ, a scalar product is defined by

$$\langle u, v \rangle_{H^\ell(\Omega)} := \sum_{|\alpha| \leq \ell} \langle D^\alpha u, D^\alpha v \rangle_{L^2(\Omega)}$$

inducing the Sobolev norm

$$\|u\|_{H^\ell(\Omega)} := \sqrt{\langle u, u \rangle_{H^\ell(\Omega)}} = \sqrt{\sum_{|\alpha| \leq \ell} \|D^\alpha u\|_{L^2(\Omega)}^2}. \tag{3.5}$$

The restriction on weak derivatives of order ℓ yields

$$|u|_{H^\ell(\Omega)} := \sqrt{\sum_{|\alpha| = \ell} \|D^\alpha u\|_{L^2(\Omega)}^2}, \tag{3.6}$$

which will be denoted as a Sobolev semi norm.

For $\ell = 0$ (3.4) implies $H^0(\Omega) = L^2(\Omega)$ and the equality of the respective scalar products. In the following, the short forms $\|\cdot\|_\ell$ and $|\cdot|_\ell$ of the norms $\|\cdot\|_{H^\ell(\Omega)}$ and semi norms $|\cdot|_{H^\ell(\Omega)}$ are used, where $\ell \geq 0$. Since $H^0(\Omega) = L^2(\Omega)$, it holds that $\|\cdot\|_0$ also denotes $\|\cdot\|_{L^2(\Omega)}$ and $\langle \cdot, \cdot \rangle_0$ the scalar product $\langle \cdot, \cdot \rangle_{L^2(\Omega)}$.

Sobolev spaces are Hilbert spaces [18]; that is, they are complete. Furthermore, the function space $C^\infty(\Omega) \cap H^\ell(\Omega)$ is dense in $H^\ell(\Omega)$, which means that each function of $H^\ell(\Omega)$ can be approximated with arbitrary precision by a function of $C^\infty(\Omega) \cap H^\ell(\Omega)$. For $C^\infty(\overline{\Omega}) \cap H^\ell(\Omega)$ this only holds with additional prerequisites on the regularity of the boundary $\partial\Omega$ [15]. The following definition will incorporate the homogeneous Dirichlet boundary conditions of the model problem (3.2) into the Sobolev spaces.

Definition 3.5 (Sobolev spaces H_0^ℓ). *For a domain $\Omega \subset \mathbb{R}^m$ the function space $H_0^\ell(\Omega)$ denotes the closure of $C_0^\infty(\Omega)$ with respect to the Sobolev norm $\|\cdot\|_{H^\ell(\Omega)}$.*

The function spaces $H_0^\ell(\Omega)$ are closed subspaces of $H^\ell(\Omega)$ and Hilbert spaces with respect to the scalar product $\langle\,\cdot\,,\,\cdot\,\rangle_\ell$. In particular, for this specific Sobolev spaces the *Poincaré Friedrich inequality*

$$\|u\|_0 \preceq |u|_1 \qquad \forall u \in H_0^1(\Omega)$$

holds. Using this inequality, it can be shown that the norms (3.5) and semi norms (3.6) are equivalent in $H_0^\ell(\Omega)$, i.e.,

$$\|u\|_\ell \asymp |u|_\ell \qquad \forall u \in H_0^\ell(\Omega), \tag{3.7}$$

with constants independent from Ω [14]. Here, $a \preceq b$ and $a \asymp b$ denote inequality and equality up to positive constants. That is, there exist real numbers $\alpha, \beta > 0$ with $a \leq \alpha b$ and $\alpha a \leq b \leq \beta a$, respectively.

The fact that Sobolev spaces are Hilbert spaces allows the analysis of the existence and uniqueness of solutions of boundary value problems by means of functional analysis, as will be described in the following section.

3.2.3 Variational formulation

For the following general description, let H denote a Hilbert space with the scalar product $\langle\,\cdot\,,\,\cdot\,\rangle_H$ and the associated norm $\|\cdot\|_H$. The set of all continuous linear functionals $l : H \to \mathbb{R}$ is denoted as *dual space H'* and equipped with the *operator norm*

$$\|l\|_{H'} := \sup_{u \in H \setminus \{0\}} \frac{|l(u)|}{\|u\|_H}.$$

In the following, the boundary value problem will be transformed into a variational problem in the Hilbert space $H \subset L^2(\Omega)$. For this purpose, the model problem (3.2) will be considered first.

Applying the L^2 scalar product with functions $v \in H_0^1(\Omega)$ to both sides of Poisson's equation $-\Delta u = f$ results in

$$\int_\Omega -\operatorname{div}(\operatorname{grad} u)v \, dx = \int_\Omega fv \, dx \qquad \forall v \in H_0^1(\Omega).$$

Green's theorem in combination with the homogeneous Dirichlet boundary conditions imply

$$\int_\Omega \operatorname{grad} u \cdot \operatorname{grad} v \, dx = \int_\Omega fv \, dx \qquad \forall v \in H_0^1(\Omega).$$

The left-hand side of this equation can be interpreted as a symmetric bilinear form $a : H \times H \to \mathbb{R}$, the right-hand side as linear functional $l : L^2(\Omega) \to \mathbb{R}$. This motivates to search the solution of the boundary value problem as solution $u \in H$ of the *variational formulation*

$$a(u, v) = l(v) \qquad \forall v \in H. \tag{3.8}$$

Such a solution u is denoted as a *weak solution* of the elliptic boundary value problem $Lu = f$ with homogeneous Dirichlet boundary condition $u = 0$.

The foundation for the existence and uniqueness of such a solution is given by the following theorem, of which the proof is given in [15].

Theorem 3.1 (Riesz representation theorem). *For each $u \in H$ a continuous linear functional $l_u \in H'$ is defined by*

$$l_u(v) := \langle u, v \rangle_H .$$

For each $l \in H'$ there exists exactly one element $u \in H$ with

$$l(v) = \langle u, v \rangle_H \qquad \forall v \in H.$$

Furthermore, it holds that $\|l\|_{H'} = \|u\|_H$.

The representation theorem suggests using the bilinear form of the variational formulation (3.8) to define the scalar product $\langle u, v \rangle_H$. Additional prerequisites are given in the following.

Definition 3.6. *The bilinear form $a : H \times H \to \mathbb{R}$ on a Hilbert space H is called* bounded, *if there is a constant $\alpha_b > 0$ such that*

$$|a(v, w)| \le \alpha_b \|v\|_H \|w\|_H \qquad \forall v, w \in H.$$

The bilinear form a is called coercive *or* elliptic *on $V \subset H$, if there exists a constant $\alpha_c > 0$ such that*

$$a(v, v) \ge \alpha_c \|v\|_H^2 \qquad \forall v \in V.$$

If the bilinear form a is symmetric, bounded, and coercive on $V \subset H$, it follows directly that

$$\alpha_c \|v\|_H^2 \le a(v, v) \le \alpha_b \|v\|_H^2 \qquad \forall v \in V. \tag{3.9}$$

This implies $a(v, v) \ge 0$ for all $v \in V$ and $a(v, v) = 0$ if and only if $v = 0$. That is, a defines a scalar product on $V \subset H$ as desired.

Lemma 3.1. *Let V be a closed subspace of the Hilbert space H and a : $H \times H \to \mathbb{R}$ a symmetric and bounded bilinear form on H, which in addition is coercive on V. This implies that V is a Hilbert space with the scalar product $a(\cdot, \cdot)$ and the associated energy norm $\|v\|_a := \sqrt{a(v, v)}$.*

Proof. It only must be shown that V is closed with respect to the energy norm $\|\cdot\|_a$. This follows directly from the equivalence of the norms

$$\|v\|_a \asymp \|v\|_H \qquad \forall v \in V$$

because of (3.9) and the fact that V is closed with respect to the norm $\|\cdot\|_H$. $\qquad\square$

The Hilbert space V equipped with the scalar product a given in Lemma 3.1 can now readily be used to derive the desired existence and uniqueness results with the aid of the Riesz representation theorem.

Theorem 3.2. *Let H be a Hilbert space, V a closed subspace of H, $l \in V'$, and $a : H \times H \to \mathbb{R}$ a bilinear form, which is symmetric and bounded on H and coercive on V. The variational problem*

$$a(u, v) = l(v) \qquad \forall v \in V$$

has a unique solution $u \in V$.

For the model problem (3.2) let $V := H_0^1(\Omega)$. It holds that

$$a(u, u) = \int_\Omega \text{grad } u \cdot \text{grad } u \, dx = \sum_{|\alpha|=1} \|D^\alpha u\|_{L^2(\Omega)}^2 = |u|_1^2.$$

Equation (3.7) implies $a(u, u) = |u|_1^2 \asymp \|u\|_1^2$. Analogous to the proof of Lemma 3.1, this results in the existence and uniqueness of a weak solution $u \in H_0^1(\Omega)$. Due to the construction of the symmetric variational problem and its unique solution, each classical solution $u \in C^2(\Omega) \cap C^0(\overline{\Omega})$ is also the solution of the variational formulation. Conversely, each weak solution $u \in C^2(\Omega) \cap C^0(\overline{\Omega})$ of the variational problem is also a classical solution of the boundary value problem [15].

In contrast to the model problem, the symmetry of the bilinear form a cannot be generally assumed. This follows directly from the counter example

$$-u'' + u' + u = f \quad \text{on } \Omega := [0, 1] \quad \text{with } u(0) = u(1) = 0$$

and the associated non-symmetric bilinear form

$$a(u, v) = \int_0^1 (-u'' + u')v \, dx = \int_0^1 u'v' \, dx + \int_0^1 u'v \, dx.$$

However, a is bounded and coercive. This will turn out as sufficient assumptions for the existence and uniqueness of a weak solution, as is stated in the following [15].

Theorem 3.3 (Lax-Milgram). *Let V be a Hilbert space, let $a : V \times V \to \mathbb{R}$ be a bounded coercive bilinear form, and $l \in V'$. The variational problem*

$$a(u, v) = l(v) \qquad \forall v \in V$$

has a unique solution $u \in V$.

The proof of Theorem 3.3 is based on deriving a contraction mapping from the variational formulation to show the existence and uniqueness of a weak solution with the aid of the Banach fixed-point theorem.

In the following, the concrete determination of an approximation of the weak solution of the variational problem is discussed, which builds the foundation of the computational aspects of the finite element method.

3.2.4 Ritz-Galerkin method

A pragmatic approach in order to determine an approximate solution of the weak solution characterized by the theorem of Lax-Milgram is to use a finite dimensional subspace $V_h \subset V$. For an arbitrary basis $\{b_1, \ldots, b_{n_h}\}$ of V_h, the approximate weak solution u_h can be written as

$$u_h = \sum_{k=1}^{n_h} c_k b_k \quad \text{with} \quad c_k \in \mathbb{R}. \tag{3.10}$$

Here, the *discretization parameter* h denotes the degree of fineness of the approximating subspace V_h of dimension $n_h \in \mathbb{N}$. That is, if h tends to zero it holds that n_h tends to infinity and $u_h \in V_h$ converges to the weak solution $u \in V$. This approach is denoted as the *Ritz-Galerkin method*.

The associated variational formulation is given by $a(u, v) = l(v)$ for all $v \in V_h$. Due to the linearity of a and l, it suffices to test all elements b_k of the basis instead of all $v \in V_h$. Using the representation (3.10) of u_h with respect to the same basis, the variational formulation can be written as

$$a\left(\sum_{k=1}^{n_h} c_k b_k, v\right) = \sum_{k=1}^{n_h} c_k\, a\,(b_k, v) = l(v) \quad \forall v \in \{b_1, \ldots, b_{n_h}\}. \tag{3.11}$$

This defines a system of linear equations for the coefficients c_k of the basis functions in the representation of the approximate weak solution u_h.

Lemma 3.2. *The coefficient vector $c = (c_k)$, $k \in \{1, \ldots, n_h\}$ of the representation (3.10) of the approximate weak solution $u_h \in V_h$ is defined as the solution of the* Ritz-Galerkin system

$$Gc = f \quad \text{with} \quad g_{k,\ell} := a(b_k, b_\ell) \quad \text{and} \quad f_k := l(b_k). \tag{3.12}$$

The Ritz-Galerkin matrix $G := (g_{k,\ell})$, $k, \ell \in \{1, \ldots, n_h\}$, *is positive definite.*

Proof. The representation of the Ritz-Galerkin system is a direct consequence of the system of equations given by (3.11). Due to the theorem of Lax-Milgram, c parameterizes the desired approximate weak solution u_h.

It remains to show that the Ritz-Galerkin matrix G is positive definite, that is, $x^t G x > 0$ holds for all $x \in \mathbb{R}^{n_h} \setminus \{0\}$:

$$x^t G x = \sum_{k=1}^{n_h} \sum_{\ell=1}^{n_h} g_{k,\ell} x_k x_\ell = \sum_{k=1}^{n_h} \sum_{\ell=1}^{n_h} a(b_k, b_\ell) x_k x_\ell$$

$$= a \left(\sum_{k=1}^{n_h} x_k b_k, \sum_{\ell=1}^{n_h} x_\ell b_\ell \right) \succeq \left\| \sum_{k=1}^{n_h} x_k b_k \right\|_0^2 > 0.$$

Here, for the estimate of the lower non-zero bound, it has been used that a is coercive and $x \neq 0$ implies $u_h \not\equiv 0$. □

Hence, the problem of solving the boundary value problem is reduced to construct a suitable subspace V_h and an appropriate set of basis functions b_k, as well as to assemble and solve the Ritz-Galerkin system. Since for $h \to 0$ the dimension of G becomes arbitrarily large, the practical solvability of the Ritz-Galerkin system has to be considered. It would be ideal if the basis functions are a-orthonormal, that is, if $a(b_k, b_\ell) = \delta_{k,\ell}$, which directly implies $c_k = f_k / g_{k,k}$. Such a basis is only available in the rarest cases. As an alternative, local trial functions with compact supports are used, which results in G being sparse, i.e., most of its entries are zero. Such compact supports or their building blocks are denoted as *finite elements*. Local basis functions are not only favorable with respect to the sparsity pattern of G, but also with respect to approximation properties like local influence of the coefficients c and flexibility in local resolution.

The question is how the choice of the approximate subspace V_h affects the quality of the approximation u_h of the weak solution u. This is answered by the following lemma.

Lemma 3.3 (Céa). *Let V_h be a finite closed subspace of the Hilbert space V, $a : V \times V \to \mathbb{R}$ be a bounded coercive bilinear form on V, and $l \in V'$. For the weak solution $u \in V$ and its approximation $u_h \in V_h$, it holds that*

$$\|u - u_h\|_V \leq \frac{\alpha_b}{\alpha_c} \min_{v_h \in V_h} \|u - v_h\|_V. \tag{3.13}$$

Here α_b denotes the constant of boundedness and α_c the constant of coercivity of a in V.

Proof. Due to the prerequisites, it holds that

$$a(u, v) = l(v) \quad \forall v \in V \quad \text{and} \quad a(u_h, v) = l(v) \quad \forall v \in V_h.$$

Subtracting these equations in V_h yields $a(u - u_h, v) = 0$ for all $v \in V_h$, and with the choice $v = v_h - u_h \in V_h$ for $v_h \in V_h$ in particular, $a(u - u_h, v_h - u_h) = 0$. Using the coercivity and boundedness of a implies

$$
\begin{aligned}
\alpha_c \|u - u_h\|_V^2 &\leq a(u - u_h, u - u_h) \\
&= a(u - u_h, u - v_h) + a(u - u_h, v_h - u_h) \\
&= a(u - u_h, u - v_h) \leq \alpha_b \|u - u_h\|_V \|u - v_h\|_V .
\end{aligned}
$$

The result follows after division by $\alpha_c \|u - u_h\|_V$ from the boundedness of V_h. $\qquad\square$

Therefore, the error $\|u - u_h\|_V$ of the approximate weak solution u_h with respect to u is asymptotically the error of the best approximation of u in V_h. In this sense, the approximation is optimal. The norm given in the estimate (3.13) of Céa's lemma results from the boundary value problem and its weak formulation. Under the assumption that Ω can be represented by a union of congruent cones, the following estimates with respect to other norms can be derived [14].

Lemma 3.4 (Aubin-Nitsche). *Let H be a Hilbert space with the scalar product $\langle \cdot, \cdot \rangle_H$ and the associated norm $\|\cdot\|_H$ and V be a closed subspace of H with respect to the norm $\|\cdot\|_V$. Then it holds for the approximate weak solution u_h in the finite subspace V_h of V*

$$
\|u - u_h\|_H \preceq \|u - u_h\|_V \sup_{g \in H} \left(\frac{1}{\|g\|_H} \inf_{v_h \in V_h} \|\varphi_g - v_h\|_V \right)
$$

if the unique weak solution $\varphi_g \in V$ of

$$
a(w, \varphi_g) = \langle g, w \rangle_H \qquad \forall w \in V
$$

is related to each $g \in H$.

In the following section, the practical aspects of the finite element approximation are described using the example of linear trial functions defined over a triangulation of the domain. In addition, the error estimates given in this section are concretized.

3.2.5   Linear trial functions over triangular elements

In the previous sections, the weak solution u has been approximated by u_h in a finite subspace V_h. The latter is defined over a tessellation of the domain Ω. To solve the model problem, this approach will be shown using the example of linear trial functions defined over a triangulation of Ω. Therefore, the domain Ω is tessellated into a conformal mesh of triangular elements according to the mesh properties stated in Section 2.2.1. The union of all mesh elements results in a polygonal domain Ω_P, which approximates Ω. To provide error estimates, additional premises must be made, which also puts the resolution parameter h of the approximate space into context.

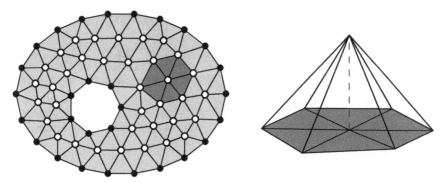

Figure 3.3: Admissible triangulation of Ω_P and linear basis function b_k.

Definition 3.7. *Let $h > 0$ denote an element size parameter. The triangulation $T_h = \{E_1, \ldots, E_{n_h}\}$ of Ω_P into $n_h \in \mathbb{N}$ closed triangular elements E_k with diameter $\leq 2h$ is called* admissible *if the union of all elements results in Ω_P, i.e., $\bigcup_{k=1}^{n_h} E_k = \overline{\Omega}_P$, and two different triangular elements E_k, E_ℓ with $E_k \cap E_\ell \neq \emptyset$ share either a common node or an edge.*

A family of admissible triangulations T_h is called quasi uniform *if there is an $\alpha_q > 0$ such that each element E of T_h contains a disc with radius $\geq h_D / \alpha_q$, where h_D denotes the halved diameter of E.*

The triangulation is called uniform, *if there is an $\alpha_u > 0$ such that each $E \in T_h$ contains a disc with radius $\geq h/\alpha_u$.*

The nodes on the polygonal boundary $\partial \Omega_P$ are called *outer nodes* or *boundary nodes*, while all other nodes are called *inner nodes*. For each inner node p_k, a basis function b_k is defined, which is linear on each triangle with the function values $b_k(p_\ell) = \delta_{k,\ell}$. That is, the value of b_k at the node p_k is one, and for all other nodes it is zero. Such a basis, which is defined by prescribed values at the nodes, is denoted as *nodal basis*. By construction, all basis functions are zero at the outer nodes, and hence on $\partial \Omega_P$, which guarantees the Dirichlet boundary conditions are met.

Figure 3.3 depicts on the left side an exemplary triangulation of the planar domain given in Figure 3.2. The inner nodes of the triangulation are marked by circles, the outer nodes are marked by discs. The piecewise linear trial function related to the support shaded dark gray is shown on the right. To fulfill the homogeneous Dirichlet boundary conditions by u_h on $\partial \Omega$, the triangulation domain Ω_P must be a subset of Ω. However, in practice usually a polygonal approximation of $\partial \Omega$ is used for triangulation. Hence, Ω_P contains points outside Ω resulting in the boundary conditions only being approximately fulfilled.

Due to the definition of the piecewise linear trial functions, the supports of b_k and b_ℓ with $k \neq \ell$ only have a common intersection with measure > 0 if their associated nodes are connected by an edge of the triangulation. As a result, the Ritz-Galerkin matrix G is sparse. Furthermore, since the partial

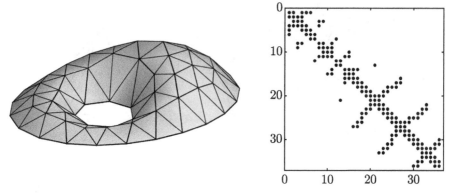

Figure 3.4: Approximate weak solution u_h and sparsity pattern of G.

derivatives of b_k are constant on each triangle, it holds that

$$g_{k,\ell} = \int_{\Omega_P} \operatorname{grad} b_k \cdot \operatorname{grad} b_\ell \, dx = \sum_{E \in T_h} \int_E \operatorname{grad} b_k \cdot \operatorname{grad} b_\ell \, dx$$

$$= \sum_{E \subset \operatorname{supp} b_k \cap \operatorname{supp} b_\ell} |E| \operatorname{const}(b_k, E) \operatorname{const}(b_\ell, E),$$

where $|E|$ denotes the area of the triangular element E. Therefore, the elements of G can easily be derived. In addition, $b_k(p_\ell) = \delta_{k,\ell}$ implies $u_h(p_\ell) = c_\ell$, i.e., the function values of u_h at each inner node p_ℓ is given by the related coefficient c_ℓ of the nodal basis. Due to the Dirichlet boundary conditions, the function values at the outer nodes are zero.

The left side of Figure 3.4 shows the approximate solution u_h of the model problem (3.2) with constant right-hand side $f \equiv 1$ using the triangulation depicted before. The sparsity pattern of the Ritz-Galerkin matrix G is depicted on the right. Here, each non-zero entry of the sparse 36×36 matrix is marked by a point. The sparsity pattern of G depends on the triangulation of Ω_P as well as on the node numbering scheme, and thus the numbering scheme of the basis functions b_k.

For a family of quasi-uniform triangulations, the following error estimates hold for the best linear approximation $P_h u$ of u [14]

$$\|u - P_h u\|_\ell \preceq h^{2-\ell} \|u\|_2, \qquad \ell \in \{0, 1\}.$$

Using the lemma of Céa, this implies the estimate $\|u - u_h\|_1 \preceq h \|u\|_2$ for the weak solution u_h of the variational problem. Under additional regularity assumptions, it holds that $\|u\|_2 \preceq \|f\|_0$. With this,

$$\|u - u_h\|_1 \preceq h \|f\|_0$$

can be derived, which gives a bound of the solution error with respect to the right-hand side of the differential equation [14]. Using the results of the Aubin-Nitsche lemma finally implies the error estimate

$$\|u - u_h\|_0 \preceq h^2 \|f\|_0.$$

Halving the discretization parameter h to improve solution accuracy and using a simple subdivision scheme for the mesh elements leads to an increase of the number n_h of basis functions by factor 4. Hence, it holds that $n_h \asymp h^{-2}$. Such a refinement of the grid width h is only sensible if the solvability of large Ritz-Galerkin systems is guaranteed. As a measure for the sensitivity of the relative error of a numerical solution of $Gc = f$, the *condition number*

$$\operatorname{cond} G := \|G\| \, \|G^{-1}\| \quad \text{with } \|G\| := \sqrt{\lambda_{\max}(G^t G)} \qquad (3.14)$$

is used [130], where $\lambda_{\max}(A)$ denotes the maximal eigenvalue of the matrix A. It has been shown in [131] that for linear trial functions and regular triangulations, it holds that

$$\operatorname{cond} G \preceq h^{-2},$$

which provides the numerical stability required for solving systems for fine discretizations. However, independent of the number of triangles used, mesh quality plays a vital role, as will be demonstrated in the next section.

3.3 Influence of mesh quality on solution accuracy

In the previous section, asymptotic estimates for the solution error have been given. However, mesh quality is hidden within the constants of the estimates. To make these clearer, this section demonstrates the influence of mesh quality on solution accuracy using the example of an analytic model problem and a specific mesh topology. Poisson's equation with homogenous Dirichlet boundary conditions

$$-\Delta u = 4 \quad \text{in } \Omega,$$
$$u = 0 \quad \text{on } \partial\Omega$$

is considered for the disc $\Omega := \{x \in \mathbb{R}^2 : \|x\| < \sqrt{3}\}$. The boundary $\partial\Omega$ of Ω is given by the circle with radius $\sqrt{3}$ and center $(0,0)$. The analytic solution is

$$u(x) := 3 - x_1^2 - x_2^2 \quad \text{for all } x = (x_1, x_2)^t \in \Omega \cup \partial\Omega.$$

This solution paraboloid u is depicted on the right of Figure 3.5 together with the domain Ω shown on the left.

An initial mesh of Ω is also depicted on the left of Figure 3.5. It consists of 19 nodes and 24 triangles. It has been generated by constructing six regular,

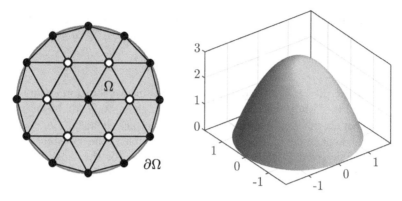

Figure 3.5: Domain and analytic solution of the model problem.

that is, equilateral triangles with edge length one around the center node $(0,0)$. On each edge of the regular hexagon defined by the six nodes marked by circles in Figure 3.5, regular triangles have again been constructed, of which the apices are located on $\partial\Omega$. Between each pair of these apices an additional node has been inserted symmetrically on $\partial\Omega$. Hence, 12 elements of the mesh are regular with mean ratio quality one and 12 elements have the mean ratio quality number 0.9386. Consequently, the mesh quality numbers are given by $q_{\min} = 0.9386$ and $q_{\mean} = 0.9693$ indicating a high quality mesh.

To demonstrate the influence of mesh quality on solution accuracy of the finite element solution with respect to the analytic solution, the initial mesh is iteratively deteriorated by rotating the six nodes marked by circles in Figure 3.5 around the center node $(0,0)$. For a given rotation angle $\alpha \in [0, \pi/3)$, the coordinates of these six nodes are given by

$$p_k^{(\alpha)} := \left(\cos(k\pi/3 + \alpha), \sin(k\pi/3 + \alpha) \right)^{\mathrm{t}}, \quad k \in \{0, \ldots, 5\},$$

with $\alpha = 0$ representing the initial mesh depicted in Figure 3.5. Using a FEM solver based on the GetFEM++ library [59], the finite element solution has been evaluated for each of the 100 meshes obtained by using the rotated nodes $p_k^{(\alpha_\ell)}$ with angles $\alpha_\ell = \ell\pi/300$, $\ell \in \{0, \ldots, 99\}$. In each case, the pointwise error $|u(x) - u_h(x)|$, $x \in \Omega \cup \partial\Omega$ obtained as the difference function of the analytic solution u and the finite element solution u_h has been determined. Results for the special choices α_ℓ with $\ell \in \{0, 33, 66, 95\}$ are depicted in Figure 3.6.

For a specific node rotation angle α_ℓ, the resulting mesh is depicted on the left. In the middle, the resulting finite element solution u_h is depicted. The absolute error function $|u - u_h|$ is shown on the right.

Iteratively rotating the inner mesh nodes distorts the outer ring of mesh elements. With this, mesh quality is decreased, as can be seen on the left. For the depicted meshes, the minimal element quality numbers $q_{\min}$ decrease from the initial value 0.9386 to 0.7885, 0.3708, and 0.0467 for $\ell = 33$, 66, and 95, respectively. In particular, for $\ell = 95$, six elements are almost degenerate,

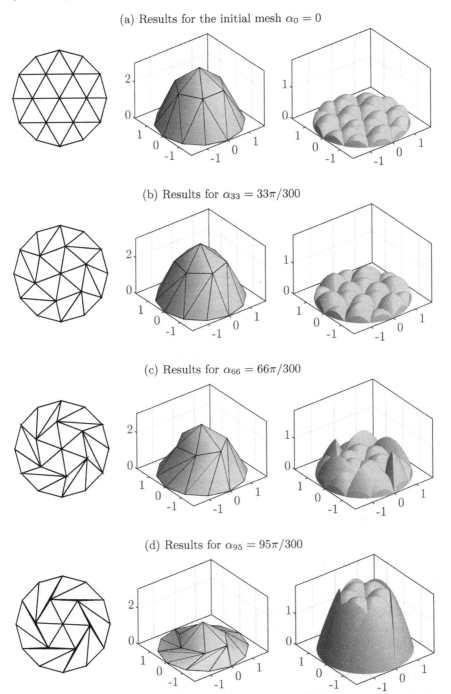

Figure 3.6: Mesh (left), FEM solution u_h (middle), and absolute error function $|u - u_h|$ (right) for different values of the node rotation parameter α_ℓ.

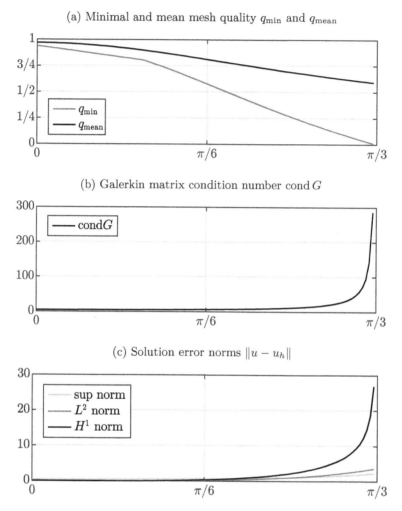

Figure 3.7: Results with respect to node rotation parameter $\alpha_\ell \in [0, \pi/3)$.

since the height of these triangles becomes almost zero. The associated mean mesh quality numbers q_{mean} decrease from 0.9693 to 0.8931, 0.7332, and 0.6074, respectively. However, mesh topology is never changed and the triangles attached to the center node are always regular.

As can be seen in the middle, rotating the inner mesh nodes leads to a flattening of the finite element solution u_h. With this, the absolute error $|u - u_h|$ depicted on the right increases significantly. Hence, good mesh quality is a fundamental prerequisite for high solution accuracy. This is also shown in Figure 3.7 depicting mesh quality numbers, Galerkin matrix condition numbers, and different error norm numbers as graphs with respect to the node rotation parameter α_ℓ.

In Figure 3.7a, the mesh quality numbers are depicted with respect to α. As can be seen, the minimal element quality number marked gray decreases continuously by increasing the rotation parameter α. For the limit angle $\alpha = \pi/3$, the apices of the lowest quality elements reach their base edges, resulting in an invalid mesh. Due to the inner elements always being regular and some outer elements with acceptable quality, the mean mesh quality q_{mean} marked black does not drop below the lower bound 0.5930, which is achieved for $\alpha_\ell = 99\pi/300$.

Since mesh topology is never changed, the sparsity pattern of the 7×7 Galerkin matrix G of the finite element approximation does not change with respect to α_ℓ. In contrast, the values of the non-zero entries depend on the nodes and with that on the rotation angle α_ℓ. This implies that the condition number $\operatorname{cond} G$ also depends on the rotation angle, as is depicted in Figure 3.7b. The condition number monotonically increases from its initial value 4.7321 for α_0 to 286.4077 for α_{99}. Although for the simple model problem considered here, the increased condition number has no significant effect on solution accuracy, it can hinder solver convergence and accuracy significantly for larger problems, as will be shown later in the example section.

Figure 3.7c depicts the solution errors with respect to the supremum norm $\|u - u_h\|_{\mathrm{sup}} := \sup_{x \in \Omega \cup \partial \Omega} |u(x) - u_h(x)|$ indicating the maximal pointwise deviation of the finite element solution, the L^2 norm $\|u - u_h\|_0$ indicating the total deviation, and the H^1 norm $\|u - u_h\|_1$ indicating the total deviation including the first derivatives. The initial errors with respect to these norms for $\alpha = 0$ are given by 0.3185, 0.0478, and 0.0887, respectively. Due to the flattening effect of the mesh distortion caused by rotating the inner nodes, the error with respect to the supremum norm increases to 2.2366 for α_{99}, which is about seven times larger if compared to the initial mesh result. The L^2 error increases by factor 76.4 to 3.6514, and the H^1 error by factor 303.4 to 26.9141. However, since linear basis functions are used, the derivatives of u_h are constant on each triangle, which only gives a rough approximation of the derivatives of u leading to large errors with respect to the H^1 norm. If derivatives of the finite element solution are of interest, higher order basis functions must be used. Nevertheless, these numbers show that solution errors can significantly increase if the mesh quality decreases. Hence, quality mesh generation and mesh quality improvement play a fundamental role in the finite element method.

4

Mesh improvement

This chapter provides an overview of mesh improvement methods and their classification. A more detailed description is given for Laplacian smoothing, which is popular due to its geometry-driven approach and its implementational and computational simplicity. In addition, a global optimization-based approach is described, which results in high mesh quality by applying mathematical optimization. Both methods are based on node relocations only and will serve as runtime and quality benchmarks in subsequent chapters. The last section discusses some exemplaric topological mesh modification techniques based on applying modification templates or node insertion and removal.

4.1 Overview and classification

As has been noted in Section 2.4, *mesh improvement methods* are a key component of the most common mesh generation techniques for the finite element method, since shape and distribution of the mesh elements have impact on the trial functions, and thus the numerical stability as well as solution accuracy [51,55,104,107,125,131,136]. Furthermore, mesh improvement methods are also applied, for example, in the process of surface reconstruction using scanner data or automated joining of different meshes [20,39,42,55,120, 127].

The various existing mesh improvement methods can roughly be classified into two distinct categories: methods that use *topological modifications*, and *mesh smoothing methods* preserving mesh topology by applying node relocations only. Typical topology-altering improvement methods refine or coarsen the mesh by element splitting or merging, node insertion or deletion, local subdivision, and edges or face swapping [43,50,55,81,93,113]. In the case of simplicial meshes, these methods usually try to achieve the Delaunay property for the mesh or try to ensure specific element properties with respect to minimum angle or edge ratios.

Mesh smoothing methods are of interest, if common mesh interfaces, boundaries, or even the complete mesh topology must be preserved, as in iterative solution methods and design optimization algorithms. Furthermore, due to the specific adjacency structure, topological modifications of hexahedral meshes are far more complex than in the case of tetrahedral meshes, since such modifications usually propagate along complete sheets. Therefore, improving hexahedral mesh quality by smoothing is much simpler than improving it by topology-altering operations.

One of the most popular smoothing methods is *Laplacian smoothing*, where each node is repeatedly replaced by the arithmetic mean of its edge-connected neighbor nodes [45, 94]. This method is popular due to its simple implementation, low numerical complexity, and fast convergence behavior. However, since Laplacian smoothing is not specifically geared towards improving element quality, possible issues of this method are deformation and shrinkage in the case of surface meshes, generation of inverted or invalid elements, and mesh quality deterioration. In particular, the method, in its basic form, is less suitable for meshes generated by adaptive octree-based methods [76] and mixed meshes, that is, meshes consisting of elements of different types, like triangles and quadrilaterals [178].

Therefore, improved methods have been proposed to circumvent these problems [164, 179]. Among these are *length-weighted Laplacian smoothing*, which uses averaging functions for neighboring nodes, and *angle-based* methods [48, 177, 178] comparing and adjusting angles incident to a node, thus reducing the risk of generating inverted elements. Alternatively, *smart or constrained Laplacian smoothing* incorporate quality aspects by applying node relocations only if they lead to an improvement of local mesh quality with respect to a given metric [21, 48]. In the case of mixed meshes, significantly better results can be obtained by using weighted means [21, 115, 177]. A more detailed discussion of Laplacian smoothing and some of its variants is given in Section 4.2.1.

Whereas geometry-based methods, like Laplacian or angle-based smoothing, determine new node positions by geometric rules, *local optimization techniques* move nodes in order to optimize objective functions based on quality numbers of the incident elements [3, 22, 170, 175]. Such methods effectively prevent the generation of degenerate or inverted elements [3, 49]. For example, [175] proposes a smoothing method by relocating interior nodes to the mass centers of their surrounding hexahedra. As in the case of smart Laplacian smoothing, the solution of this minimization problem is only applied if quality is improved. A node quality metric-based local optimization method using a combined gradient-driven and simulated annealing technique-based approach is presented in [19].

Due to the incorporation of element quality metrics and standard optimization techniques, like the conjugate gradient method or linear programming, the computational costs of local optimization-based techniques are increased if compared to geometry-based methods. Therefore, *combined methods* have been proposed using variants of Laplacian smoothing as well as local optimization techniques. Here, local optimization is only applied in the case of elements with a quality below a given threshold in order to find a balance between quality and computational effort [21, 48, 90].

Whereas local optimization techniques are well suited in order to efficiently resolve mesh problems locally, *global optimization-based smoothing* methods incorporating all mesh elements are geared towards effectively improving the overall mesh quality [8, 16, 53, 87, 117, 128, 173]. Due to their holistic approach, global optimization techniques result in a superior overall mesh quality. However, depending on the quality metrics and optimization techniques involved, these methods can be very demanding in terms of computation time, memory, and implementational complexity. This can be circumvented to some extent by using streaming techniques [169]. A more in-depth description of a global optimization-based mesh smoothing approach and its main components is given in Section 4.2.2.

Specialized optimization-based methods exist for surface meshes to preserve the given shape and its characteristics, such as discrete normals and curvature [23, 44, 57]. They are mainly based on using local parameterizations and modified objective functions. A combined global optimization approach improving element size as well as shape is presented in [128]. Here, smoothing is driven by minimizing Riemannian metric non-conformity. Since the objective function is not easily differentiable, a brute force approach is used for minimization.

All optimization-based methods have in common that the proper choice of quality metrics and objective functions plays a crucial role [78, 84, 85]. For example, in the case of mixed mesh smoothing, quality metrics must be applicable and balanced in their behavior for all element types under consideration. In addition, optimization-specific additional requirements like differentiability, or practice-oriented requirements like evaluation efficiency, should be considered. Hence, a lot of effort has been put into finding suitable quality metrics.

Common quality metrics are based on geometric entities like angles, lengths, and surface areas, as well as their aspect ratios [55]. However, in the context of mesh optimization, algebraic quality metrics are more suitable [28,52,84,85,100]. Furthermore, by a proper choice of the objective function and quality metrics, global optimization can also be used to untangle meshes [83]. In this context, [86] provides a more generalized approach to the construction of appropriate and flexible objective functions with respect to optimization aims, by introducing the target-matrix paradigm. Here, for each sample point of the mesh, two matrices are required. One is the Jacobian matrix of the current mesh, the other a target-matrix representing the desired Jacobian matrix in the optimal mesh. Usually, sample points are given by element nodes and the target matrices are derived from type-dependent reference elements. Using objective functions based on these matrices allows mesh quality to be defined on a higher conceptual level. An example for this approach is the mean ratio criterion, for which the definition is given in Section 2.3.2. By analyzing two energy functions based on conformal and isoparametric mappings, this metric was shown in [78] to be equivalent to the angle-preserving energy. Hence, mesh optimization results in minimizing the energy.

Due to the rapidly growing complexity of simulations nowadays, there is a great demand for fast mesh improvement methods providing high quality results. In this context, the computational effort of global optimization-based methods may become computationally too expensive. As an alternative, the geometric element transformation method (GETMe) has been proposed [141,143,153,156, 159,160,163], which is also the main subject of this book. Instead of improving element shape based on solving optimization problems, GETMe achieves mesh improvement by applying specific geometric element transformations, which successively transform an arbitrary mesh element into its regular counterpart [151,155,158,160]. This is optionally combined with a relaxation and weighted node averaging scheme involving mesh quality. In doing so, GETMe smoothing combines a global optimization-like smoothing effectivity with a Laplacian smoothing-like runtime efficiency.

Further specialized mesh smoothing approaches exist. For example, methods based on physical models like systems of springs [95], packings of bubbles [126], and solving a global system derived from local stiffness matrices [8] or statistical approaches [41]. In [98] the mesh is considered as a deformable solid. In doing so, smoothing of the mesh is treated as a matrix problem in finite elements. Alternatively, [118] proposes a method for improving hexahedral meshes by quasi-statistical modeling of mesh quality parameters, and the approach given in [119] is based on space mapping techniques. In [65], mesh smoothing is conducted by solving a large nonlinear algebraic system obtained by Laplace-Beltrami equations for an unstructured mixed element volume mesh. However, solving the system by using Newton-Krylov methods is a critical task, as issues in solver convergence observed in numerical examples have shown.

There also exist various smoothing methods that are more intimately connected with the finite element approximation process; these include the a priori-based approach given in [33]. Here, optimization is based on a measure of the interpolation error associated with the finite element model. In contrast, a posteriori smoothing methods incorporate the information obtained by error estimates [2,9,106]. Such adaptive techniques usually apply mesh improvement and refinement within a finite element iteration scheme. They are particularly well suited for anisotropic problems or problems with singularities.

Each of the existing methods for mesh improvement faces some distinct problems. The topology-based methods often rely heavily on the insertion of new elements or the splitting of old ones to improve mesh quality. In this case, the resulting mesh might be comprised of an excessive number of elements or elements of very small dimensions. Thus, numerical solvers typically used for FEM computations can be severely affected due to the need for unreasonable computing power, or due to the ill-conditioned numerical models arising when miniscule elements exist. Furthermore, altering the number of elements makes the mapping of results in similar models cumbersome. On the other hand, smoothing methods are restricted by the given mesh topology, since adverse topological configurations prevent smoothing methods from reaching high quality results. Thus it is natural to develop *hybrid methods* combining topological modifications with smoothing and optimization methods as proposed in [21, 48]. In these hybrids, each method can be used as a preconditioner for the other, or iteratively until the desired mesh quality is achieved.

4.2 Mesh smoothing

4.2.1 Laplacian smoothing and its variants

Due to its simplicity and efficiency, one of the most popular smoothing methods until today is the *Laplacian smoothing* approach. Let p_i denote an arbitrary node of the mesh and $J(i)$ the index set of all other mesh nodes, which are connected to p_i by an edge. The new position p_i' of p_i is derived as

$$p_i' := \frac{1}{|J(i)|} \sum_{j \in J(i)} p_j, \tag{4.1}$$

where $|J(i)|$ denotes the number of elements in the index set $J(i)$. As can be seen, p_i' is the arithmetic mean of its edge-connected nodes [45, 94]. All nodes are updated according to (4.1) and this is iterated until the maximal node movement distance $\max_i \|p_i' - p_i\|$ drops below a given threshold. In the case of a rectangular grid, this method can be derived by the finite difference approximation of the Laplace operator, which also justifies its name. Iteratively updating the nodes in Laplacian smoothing affects subsequent computations and implies that results depend on the order of the nodes. Furthermore, since no quality or validity criteria are incorporated, mesh quality can deteriorate and invalid elements can be generated.

A variant is the *area- or volume-weighted Laplacian smoothing* approach given in [79]. For an arbitrary mesh element E_j, let c_j denote its centroid obtained as the arithmetic mean of all nodes of E_j, and v_j the area or volume of E_j in the 2D or 3D case, respectively. Within one smoothing loop, each interior node p_i is corrected by the movement vector

$$\Delta p_i := \frac{\sum_{j \in \tilde{J}(i)} v_j (c_j - p_i)}{\sum_{j \in \tilde{J}(i)} v_j}, \tag{4.2}$$

where $\tilde{J}(i)$ denotes the index set of all elements attached to the node p_i. This is repeated until the maximal absolute movement $\max_i \|\Delta p_i\|$ drops below a given threshold. As in the case of Laplacian smoothing by successively updating all nodes, results of this approach depend on the node order. Furthermore, due to (4.2) and the geometric termination criterion, area- or volume-weighted Laplace does not involve quality measures and thus cannot guarantee quality improvements and the validity of the resulting mesh.

These issues are addressed by the *smart Laplacian smoothing* approach

given in [48]. It is based on the classical Laplacian smoothing scheme, where
the new position of an interior node is derived as the arithmetic mean of all
edge-connected nodes. However, this new node position is only set if this leads
to an increase of the arithmetic mean of quality values of all attached elements.
Evaluating these local mean quality values before and after node movement
results in a significantly increased computational effort.

To ensure that the results of smart Laplacian smoothing are independent
of the node order, which is relevant if run in parallel, the following additional
modification is also applied. Quality-improving node updates are not conducted
immediately, but stored separately until all nodes have been tested. After that,
all relevant node updates are applied. This may lead to the generation of
invalid elements. Therefore, nodes with incident invalid elements are reset to
their previous position. This is iterated until no invalid element remains in the
mesh. Smoothing is terminated if the improvement of the arithmetic mean of
all element quality numbers drops below a given tolerance.

In [93] Lo describes the *QL smoothing method* for triangular meshes, which
is also a modified version of Laplacian smoothing. It is based on the α-quality
number of a triangle E with nodes p_0, p_1, and $p_2 \in \mathbb{R}^2$ given as

$$\alpha(E) := \frac{2\sqrt{3}(p_1 - p_0) \times (p_2 - p_0)}{\|p_1 - p_0\|^2 + \|p_2 - p_1\|^2 + \|p_0 - p_2\|^2}.$$

Here, $(p_1 - p_0) \times (p_2 - p_0)$ represents the signed area of the triangle. Hence
it holds that $\alpha(E) < 0$, if E is inverted. For an inner mesh node p_i let $\tilde{J}(i)$
denote the index set of surrounding triangles E_j, $j \in \tilde{J}(i)$. The geometric mean
α-quality of this surrounding triangles is given by

$$\overline{\alpha}(p_i) := \left(\prod_{j \in \tilde{J}(i)} \alpha(E_j) \right)^{1/|\tilde{J}(i)|}.$$

The basic idea behind QL smoothing is not to simply set the new position
of p_i to p_i' according to (4.1), but to use a suitable point on the line through
p_i and p_i' defined by $p_i^{(\lambda)} := (1 - \lambda)p_i + \lambda p_i'$, with $p_i^{(0)} = p_i$ and $p_i^{(1)} = p_i'$. The
criteria for choosing λ is based on maximizing $\overline{\alpha}(p_i^{(\lambda)})$. However, in order not
to apply a costly optimization method requiring many $\overline{\alpha}(p_i^{(\lambda)})$ evaluations
for different values of λ, only a discrete set of λ values is tested. Here, Lo
proposes to use the values 0.9, 1.0, and 1.1 for λ. That is, next to the standard
position p_i', two alternative positions are tested, representing an over- and
under-relaxation approach. This enhanced Laplacian smoothing scheme based
on the quality assessment at some strategic points using the α-quality number
is denoted as QL smoothing. Although the QL method is proposed for the
α-quality number, it can similarly be defined for any other suitable quality
criterion.

4.2.2 Global optimization

Whereas Laplacian smoothing variants are easy to implement and comparably fast, the resulting mesh quality generally cannot compete with that obtained by *global optimization-based methods* [21, 53]. These methods improve overall mesh quality by minimizing an objective function incorporating quality numbers of all mesh elements. Optimization is conducted by using standard methods, like the conjugate gradient approach, Newton's algorithm, or an active set algorithm.

As a representative of this class, the shape improvement wrapper of the mesh quality improvement toolkit *Mesquite* [99] is described below. This state-of-the-art global optimization-based approach will also serve as a quality benchmark in subsequent examples. Smoothing is based on minimizing an objective function representing the arithmetic mean of the quality of all mesh elements, and minimization is accomplished by using a feasible Newton approach. In the following, the key components of the instruction queue of Mesquite used by the shape improvement wrapper are described in more detail, mainly following the description given in [16]:

Quality metric: The quality metric used by the shape improvement wrapper is based on the inverse of the mean ratio quality criterion described in Section 2.3.2. For each element E of the mesh, the mean ratio quality number $q(E)$ represents the deviation of an arbitrary valid element from its regular counterpart. In addition to the quality metric value, the analytic gradient and Hessian information is also provided, which is required by the optimization algorithm.

Objective function: While the quality metric provides a way to evaluate the properties of individual mesh entities, the objective function provides a way of combining those values into a single number for the entire mesh. In addition, the gradient and Hessian with respect to the vertex positions are determined. Following a global optimization-based approach, in the shape improvement wrapper of Mesquite, the arithmetic mean of the inverse of the mean ratio number $q(E_i)$ of all mesh elements E_i is used as an objective function, i.e.,

$$\sum_{i=1}^{n_E} \frac{1}{q(E_i)} \;\to\; \min$$

is determined, where n_E denotes the number of all mesh elements.

Quality improver: This method takes as input the objective function and makes extensive use of the gradient and Hessian information provided therein. In addition, a termination criterion must be provided to stop iterating over the mesh. As vertex mover, the feasible Newton algorithm is applied. Newton's method minimizes a quadratic approximation of a non-linear objective function and is known to converge super-linearly near a non-singular local minimum. Therefore, mesh-optimization problems that are performed within the neighborhood of the minimum are favorable for this approach. The algorithm requires the objective function value, gradient, and Hessian. The latter is sparse for the objective function described above, which allows all vertex positions to be improved simultaneously. As a drawback, mesh configurations not near a local minimum may require a significant computational effort to be smoothed, as will be shown by distorted mesh examples given in Chapter 7.

Termination criterion: This criterion controls the termination of the smoothing process. In case of the shape improvement wrapper, termination is based on the L^2-norm of the gradient.

Additional components involved are the quality assessor, which evaluates the quality metric for all mesh elements and accumulates statistical information, and a mesh-untangling preprocessing step to ensure mesh validity, which is a condition in the definition of the mean ratio quality criterion.

4.3   Topological modifications

Since mesh smoothing is based on node relocation only, smoothing cannot overcome topological deficiencies of the mesh. However, element quality numbers resulting from smoothing algorithms can be used as an indicator for the need of topological modifications. These can, for example, be based on applying specific templates for local adjacency modification, node insertion, or removal combined with local remeshing. Such approaches will be described in the following.

4.3.1   Template-based improvement

The *template-based topological mesh improvement* approach changes specific adjacency structures of a small number of elements into another specific configuration with better quality. Here, templates depend on the element type. Some of these templates, given, for example, in [93], are depicted in Figure 4.1. In this, each element is shaded according to its mean ratio quality number with a quality grayscale bar given at the bottom of the figure. For simplicity reasons and better visibility of element quality, only templates for planar elements have been visualized. However, similar templates exist for volumetric elements.

Figure 4.1a depicts a simple *edge swap* approach for two neighboring elements. As depicted on the left, a common edge of two triangles is swapped to the other diagonal of the quadrilateral hull polygon if element quality is improved. Here, *hull polygon* denotes the boundary of the polygonal domain obtained by the union of all template elements.

Criteria for edge swapping can be based, for example, on improving the minimal quality number of the involved elements or improving their mean quality number. Whereas for triangular elements the edge swap is unique, one must choose between two possible new edges in the case of quadrilateral elements depicted on the right of Figure 4.1a. However, the same element

(a) Edge swap

(b) Node removal

(c) Short edge removal

(d) Small element removal

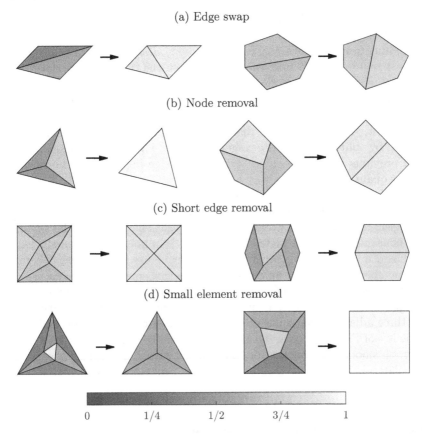

0   1/4   1/2   3/4   1

Figure 4.1: Templates for topological modifications improving mesh quality. Elements are shaded according to their mean ratio quality number $q(E) \in [0, 1]$.

quality-based criteria for the new resulting elements can be used for the choice between new edge candidates.

In contrast to edge swapping, the number of mesh elements is reduced by the *node removal* approach depicted in Figure 4.1b. In a triangular or quadrilateral mesh, the ideal number of elements adjacent to an inner node amounts to six or four, respectively. If the number of adjacent elements is significantly below this number, mesh quality can be improved by removing this node and splitting the remaining hull polygon as required. For triangular meshes and a node with only three adjacent elements, this is depicted on the left of Figure 4.1b. Here, the inner node can simply be removed and the remaining hull polygon is already a new triangular element. In the case of four triangles adjacent to the node, the remaining hull polygon is a quadrilateral, which must be split by one of its diagonals into two triangles. The case of three quadrilaterals adjacent to a common node is depicted on the right of Figure 4.1b. As in the case of the node removal for four triangles, the hull polygon must be split by one additional edge to result in two better-quality quadrilaterals.

Modification templates based on *removing a short edge* are depicted in Figure 4.1c. On the left side, the edge between the two non-boundary nodes is contracted to a single node. In consequence, the two adjacent triangular elements are removed, since they each collapse to an edge. Hence, for this configuration, the number of elements decreases from six to four. As depicted on the right, short edges inside the hull hexagon of four quadrilaterals can be removed by splitting the hull hexagon by a new edge into two quadrilaterals. As in previous cases, the choice of the new edge depends on the resulting element qualities.

For higher order polygonal elements, template-based modification is advantageous, since the hull polygon remains unchanged, which ensures that topological modifications have only a local effect. This holds for the *small element removal* templates depicted in Figure 4.1d. In both cases, a small element is enclosed by a hull polygon, which can either be used directly or split into a fewer number of elements. As depicted on the left, small triangles enclosed by an outer triangle can also be contracted to a common node. By doing so, this template converts seven elements to three elements. For the resulting node with three adjacent triangular elements, the node-removal template depicted on the left of Figure 4.1b can be applied. In practice, topological modifications and mesh smoothing are therefore applied iteratively. The replacement of five quadrilateral elements by their quadrilateral hull element is depicted on the right of Figure 4.1d.

4.3.2   Node insertion and removal

The template modifications presented in the previous section modify specific topological configurations based on element quality numbers. In contrast, *node insertion and removal* approaches modify mesh topology based on mesh density and element quality by local remeshing. For simplicial meshes, such approaches can easily be based on the same techniques as used in iterative Delaunay mesh generation methods, using the empty sphere criterion for applying local changes.

Things become more complicated in the case of polyhedral elements with a larger number of element nodes, since topological modifications tend to have a global effect. This can already be seen in the case of quadrilateral elements. If, for example, one element of a regular quadrilateral grid must be split in half by introducing new nodes on the midpoints of opposite element edges, this directly affects neighboring elements. Hence, to preserve good mesh quality, such modifications may propagate through a complete layer of the mesh until the mesh boundary is reached. Due to these difficulties, topological modifications are less common for higher order elements.

5

Regularizing element transformations

In this chapter, regularizing geometric transformations for polygonal elements with any number of nodes and the most common polyhedral finite elements are presented. If applied iteratively, these transformations convert arbitrarily shaped elements into their regular counterparts. A mathematical proof for the regularizing effect of the polygonal transformations leads to a full classification with respect to both their parameters and the resulting limit shapes. Although primarily focusing on regular limits, a scheme to derive custom transformations for prescribed non-regular limit polygons is also given. For polyhedral elements, evidence for the regularizing property is given by numerical tests. The regularizing transformations obtained in both cases will be the driving force of the GETMe algorithms for mesh smoothing presented in the subsequent chapter.

5.1 Transformation of polygonal elements

5.1.1 Classic polygon transformations

Laplacian smoothing on the one hand and global optimization-based approaches on the other somehow represent the poles of the smoothing world. Whereas Laplacian smoothing is a simple geometry-based and computationally inexpensive scheme, it is not quality driven. In contrast, global optimization approaches use the full mathematical power of numerical optimization to focus entirely on mesh quality in terms of the involved quality criterion and target function. This comes at the expense of a significantly increased computational complexity.

The question is whether there is a lightweight geometric smoothing method naturally tending to generate regular elements without incorporating quality measures. Thinking of single triangular elements, there is a well-known geometric approach to transforming arbitrary triangles into regular, i.e., equilateral ones. It is called *Napoleon's theorem*, and states that the centroids of the equilateral triangles constructed on the sides of any triangle are themselves the nodes of an equilateral triangle. This is depicted in Figure 5.1 with the initial polygon marked gray, the equilateral triangles constructed on the sides of this triangle marked by gray dashed lines, and the resulting regular Napoleon triangle marked black.

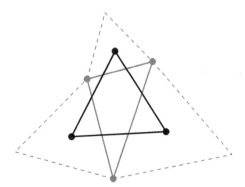

Figure 5.1: Construction of the Napoleon triangle.

An alternative construction scheme to transform arbitrary triangular elements within one transformation step into regular triangles has been proposed in [142]. Even polygonal elements with an arbitrary number $n \geq 3$ of nodes can be geometrically transformed by $n-2$ steps into their regular counterparts. This is the central result of the *Petr-Douglas-Neumann theorem* [34,101,108]. It is based on constructing isosceles triangles with base angle $\theta_k = \pi(n-2k)/(2n)$ on the sides of the polygon. Connecting the apices of these isosceles triangles results in a new polygon, which is transformed using the next angle θ_k, where $k \in \{1,\ldots,n-2\}$. Furthermore, applying any permutation of these $n-2$ transformations leads to a regular n-gon.

The simplest case, $n = 3$, resulting in $\theta_1 = \pi/6$, represents Napoleon's theorem as depicted in Figure 5.1. Here, equilateral triangles are constructed on the sides of an initial triangle marked gray. Using the centroids of these equilateral triangles as apices of triangles constructed on the same sides results in isosceles triangles with base angle θ_1.

For $n = 4$, the two transformations with base angles $\theta_1 = \pi/4$ and $\theta_2 = 0$ must be applied. This is depicted in the upper row of Figure 5.2. The left side of the upper row shows the initial polygon marked gray and the first iterate marked black, derived by constructing isosceles triangles marked by gray dashed lines with base angle θ_1 on each side. The right figure shows the construction for θ_2 based on the resulting polygon of the first transformation step. Due to $\theta_2 = 0$, in this case the isosceles triangles are of height zero, thus the apices are the midpoints of the edges.

The lower row of Figure 5.2 depicts an example for $n = 5$. Here, the transformations for $\theta_2 = \pi/10$, $\theta_1 = 3\pi/10$, and $\theta_3 = -\pi/10$ have been applied from left to right, resulting in the expected regular pentagon.

In both cases, the first node of the initial polygon and the first node of the polygon obtained by applying a transformation step are marked by circles. As can be seen by the position of these nodes, the given transformations lead to a rotation of the polygon. Furthermore, only $n-2$ transformation steps are required to regularize an element with n nodes. Particularly for the most

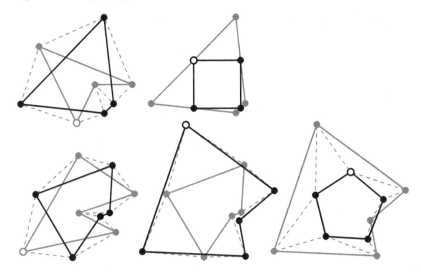

Figure 5.2: Petr-Douglas-Neumann construction steps for deriving a regular quadrilateral (upper) and regular pentagon (lower). Circle markers indicate the first node of a polygon.

common elements, the triangle and the quadrilateral, this implies a rapid change of shape.

In an iterative mesh smoothing context, individual element improvements must be orchestrated in order to lead to an improvement of the entire mesh. From this point of view both transformation properties are disadvantageous, since they increase the risk of generating inverted elements. However, both can easily be overcome by a proper choice of the transformation and its parameters as is described in the following section. In this section, a transformation scheme preserving the orientation while being controllable with respect to its regularizing effect will be given.

Figure 5.3 depicts exemplarily the effect of iteratively applying such more suitable transformations. Each row shows from left to right the initial polygon and the polygons resulting from iteratively applying a *regularizing transformation*. Here, a transformation is called regularizing if iteratively applying it to an arbitrary initial element leads to a sequence of elements tending to a regular limit element. In all cases an additional scaling step has been applied after each transformation step to preserve the sum of all edge lengths. Such a scaling scheme is denoted as *edge length sum-preserving scaling*. As in the case of the Petr-Douglas-Neumann transformation, the first node of each polygon is marked by a circle. For comparison reasons, the initial polygons depicted on the left are the same as those used for illustrating the Napoleon and Petr-Douglas-Neumann transformations in Figure 5.1 and Figure 5.2.

As can be seen, the shapes change more gradually and iteratively applying the regularizing transformation does not lead to a rotational effect, as is

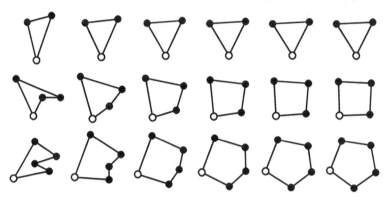

Figure 5.3: Element regularization by iteratively applying a regularizing transformation from left to right. Circles indicate the first node of a polygon.

indicated by the first node marked by a circle. Intending to improve only the worst element of a mesh by transforming it, such properties are favorable since worst-element node modifications inevitably have an impact on neighboring elements.

5.1.2 Generalized regularizing polygon transformations

5.1.2.1 Construction of regularizing polygon transformations

New types of well-suited regularizing transformations have been proposed by the authors in [156, 158] to overcome the two mayor drawbacks of the Petr-Douglas-Neumann construction. Instead of using different base angles θ_k per transformation step, the same angle $\theta \in (0, \pi/2)$ is applied in each step. This results in a more gradual and controllable change of shape during mesh smoothing, as is illustrated by Figure 5.3. In addition, the rotational effect of the transformation is eliminated by applying a combination of two oppositely directed transformation steps. Furthermore, the transformation scheme is generalized by using similar triangles instead of isosceles triangles, resulting in an additional degree of freedom.

As will turn out later, for the analysis of the resulting planar polygon transformation scheme, it is favorable to use a representation by complex numbers. That is, the node $p_k = (x_k, y_k)^t \in \mathbb{R}^2$ will be represented by the complex number $z_k = x_k + iy_k \in \mathbb{C}$, where $i := \sqrt{-1}$ denotes the imaginary unit. A representation of the regularizing transformation for nodes p_k with real number coordinates will be given by Lemma 5.12 in Section 5.1.2.3 after the complex-valued analysis.

A polygon with $n \geq 3$ nodes is defined by the vector $z := (z_0, \ldots, z_{n-1})^t \in \mathbb{C}^n$. Its n edges are $e(z_k, z_{k+1})$, with $k \in \{0, \ldots, n-1\}$. Here, all indices must be taken modulo n. For example, in the case $k = n - 1$, the expression $e(z_{n-1}, z_n)$ denotes the edge $e(z_{n-1}, z_0)$. Furthermore, all edges are oriented

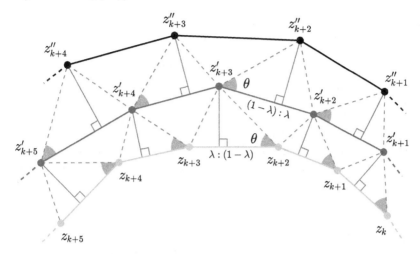

Figure 5.4: Part of an initial polygon z (light gray), polygon z' (dark gray) after the first transformation, and polygon z'' (black) after the second transformation with opposite orientation. Both use $\lambda = 1/3$ and $\theta = \pi/4$.

according to the order of the nodes in the polygon given by the vector z. That is, z_k is the start point of $e(z_k, z_{k+1})$, and z_{k+1} is its end point.

The first substep of the transformation is based on constructing an equally oriented similar triangle on each edge of the polygon. Taking the apices of these triangles results in a new polygon that likewise has n nodes. The shape of the similar triangles can be uniquely defined by two parameters λ and θ, as depicted in Figure 5.4. In this a part of an initial polygon with nodes z_k is shown marked light gray, and the triangles constructed on each edge $e(z_k, z_{k+1})$ are marked by gray dashed lines.

Here $\lambda \in (0, 1)$ represents a fixed subdivision parameter. At each subdivision point $\lambda z_k + (1 - \lambda)z_{k+1}$, located on the edge $e(z_k, z_{k+1})$, a perpendicular is constructed to the right of the directed edge. On this, the new apex z'_{k+1} is chosen in such a way that $e(z_k, z'_{k+1})$ and $e(z_k, z_{k+1})$ enclose a predefined angle $\theta \in (0, \pi/2)$. The height of the triangle is given by $(1 - \lambda)|z_{k+1} - z_k| \tan \theta$, and the normalized direction of the perpendicular is $-\mathrm{i}\,(z_{k+1} - z_k)\,/\,|z_{k+1} - z_k|$, since multiplication by $-\mathrm{i}$ represents a clockwise rotation by $\pi/2$. Thus, the apex can be written as the sum of the subdivision point and the perpendicular direction multiplied by the height, resulting in

$$
\begin{aligned}
z'_{k+1} &:= \left(\lambda z_k + (1 - \lambda)z_{k+1}\right) - \mathrm{i}(1 - \lambda)\tan\theta\,(z_{k+1} - z_k) \\
&= \underbrace{\left(\lambda + \mathrm{i}(1 - \lambda)\tan\theta\right)}_{=:w} z_k + \underbrace{\left((1 - \lambda) - \mathrm{i}(1 - \lambda)\tan\theta\right)}_{=1-w} z_{k+1} \\
&= w z_k + (1 - w)z_{k+1}.
\end{aligned}
\tag{5.1}
$$

As can be seen in (5.1), the apex z'_{k+1} is a linear combination of the nodes of the associated edge $e(z_k, z_{k+1})$ using the complex weight

$$w := \lambda + \mathrm{i}(1 - \lambda)\tan\theta,$$

which depends on the construction parameters λ and θ.

The apices z'_k and the associated polygon $z' := (z'_0, \ldots, z'_{n-1})^{\mathrm{t}}$ marked dark gray in Figure 5.4 build the basis for the second substep of the transformation. This uses an opposite construction direction to compensate the rotational effect of the first substep. In doing so, the factors λ and $(1 - \lambda)$ in the representation of the perpendicular are interchanged and the angle θ is measured at the end point of the edge instead of the start point. It follows that the apices z''_k are given by

$$
\begin{aligned}
z''_k &= \Big((1 - \lambda)z'_k + \lambda z'_{k+1}\Big) - \mathrm{i}(1 - \lambda)\tan\theta\left(z'_{k+1} - z'_k\right) \\
&= \Big((1 - \lambda) + \mathrm{i}(1 - \lambda)\tan\theta\Big)z'_k + \Big(\lambda - \mathrm{i}(1 - \lambda)\tan\theta\Big)z'_{k+1} \\
&= (1 - \overline{w})z'_k + \overline{w}z'_{k+1},
\end{aligned}
\tag{5.2}
$$

with $\overline{w}$ denoting the complex conjugate of w. Again, the apices z''_k define a polygon $z'' = (z''_0, \ldots, z''_{n-1})^{\mathrm{t}}$, which is the final transformation result depicted in black in Figure 5.4.

Due to the representations (5.1) and (5.2) that show linear combinations of nodes, these polygon transformations can conveniently be written as matrix multiplications. That is, $z' = M'z$ and $z'' = M''z'$ where the entries of the $n \times n$ matrices M' and M'' are given by

$$
M'_{j,k} := \begin{cases} 1 - w & \text{if } j = k \\ w & \text{if } j = k + 1 \\ 0 & \text{otherwise} \end{cases}
\quad \text{and} \quad
M''_{j,k} := \begin{cases} 1 - \overline{w} & \text{if } k = j \\ \overline{w} & \text{if } k = j + 1 \\ 0 & \text{otherwise,} \end{cases}
\tag{5.3}
$$

respectively, for $j, k \in \{0, \ldots, n - 1\}$. It should be noticed that zero-based indices have been used again and all indices must be taken modulo n. Due to this construction, M'' is the conjugate transpose of M' and vice versa, that is, $M'' = (M')^*$ holds. The complex weight w depends on the transformation parameters λ and θ, which also holds for the transformation matrices M' and M''. Furthermore, both matrices are circulant. That is, the entries of each matrix row can be obtained by a circular shift to the right of the entries of its preceding row.

Multiplying these matrices of the two transformation substeps conveniently results in the combined transformation $z'' = Mz$, with $M := M''M'$. It directly maps the initial polygon z marked light gray in Figure 5.4 on the final transformed polygon z'' marked black. Since the circulant matrices M' and M'' commute, that is, $M = M''M' = M'M''$, the order of applying the sub-transformations is irrelevant. Furthermore, M' and M'' are adjoint, which implies that M is a circulant Hermitian matrix. Using that $w\overline{w} = |w|^2$ in the representation of M results in the following definition.

Definition 5.1 (Generalized polygon transformation). *For $\theta \in (0, \pi/2)$, $\lambda \in$ $(0,1)$, and*

$$w := \lambda + i(1 - \lambda)\tan\theta$$

the generalized polygon transformation is given by the circulant Hermitian matrix M with entries defined by

$$M_{j,k} := \begin{cases} |w|^2 + |1 - w|^2 & \text{if } j = k \\ w(1 - \overline{w}) & \text{if } j = k + 1 \\ \overline{w}(1 - w) & \text{if } k = j + 1 \\ 0 & \text{otherwise,} \end{cases} \tag{5.4}$$

where $j, k \in \{0, \ldots, n - 1\}$. The polygon $z^{(\ell)}$ obtained by applying the transformation $\ell \geq 0$ times is given by $z^{(\ell)} := M^{\ell} z^{(0)}$, where $z^{(0)} \in \mathbb{C}^n$ denotes an initial polygon.

For triangular elements, i.e., $n = 3$, the transformation matrix (5.4) evaluates to

$$M = \begin{pmatrix} |w|^2 + |1 - w|^2 & \overline{w}(1 - w) & w(1 - \overline{w}) \\ w(1 - \overline{w}) & |w|^2 + |1 - w|^2 & \overline{w}(1 - w) \\ \overline{w}(1 - w) & w(1 - \overline{w}) & |w|^2 + |1 - w|^2 \end{pmatrix}, \tag{5.5}$$

and for quadrilateral elements, i.e., $n = 4$, the transformation is given by

$$M = \begin{pmatrix} |w|^2 + |1 - w|^2 & \overline{w}(1 - w) & 0 & w(1 - \overline{w}) \\ w(1 - \overline{w}) & |w|^2 + |1 - w|^2 & \overline{w}(1 - w) & 0 \\ 0 & w(1 - \overline{w}) & |w|^2 + |1 - w|^2 & \overline{w}(1 - w) \\ \overline{w}(1 - w) & 0 & w(1 - \overline{w}) & |w|^2 + |1 - w|^2 \end{pmatrix}. \tag{5.6}$$

These representations of M and that for $n = 5$ have been used to iteratively transform the initial random polygons $z^{(0)}$ depicted on the left of Figure 5.5. The subsequent columns show the resulting polygons $z^{(\ell)}$ after $\ell \in \{1, 2, 3, 5, 20\}$ iteratively applied transformations using $\lambda = 1/2$ and the angle θ given in front of each row. Since the outward directed construction scheme enlarges the polygon by each transformation step, all polygons have been scaled for a better representation.

As can be seen, in the first three rows, the transformation of a triangle, quadrilateral, and pentagon using the given transformation parameters leads to the desired regularization. In contrast, the last row, although using the same initial pentagon as in the row above, results in a star shape. That is, the transformation parameters have an impact on the shape of the resulting polygon. As will turn out later, only specific parameter areas lead to the desired regular polygons, whereas other choices of the parameters λ and θ lead to other kinds of symmetric configurations.

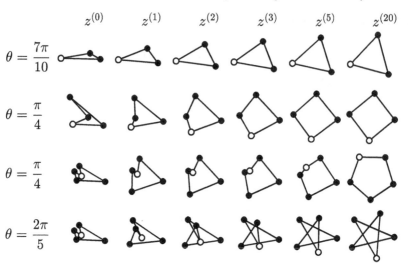

Figure 5.5: Polygons obtained by iteratively applying the generalized polygon transformation using $\lambda = 1/2$ and the given angle θ. Circles indicate the first node of a polygon.

5.1.2.2   Transformation properties

In this section, properties of the generalized polygon transformation will be analyzed. Readers not interested in the mathematical details may skip the following and continue with reading the central result given by Theorem 5.1 to find proper parameters for the transformation matrix to result in regular polygons. For those interested in a deeper insight into the properties of the regularizing polygon transformations, the preservation of the element centroid is shown first.

Definition 5.2 (Polygon centroid). *The* centroid $c(z)$ *of a polygonal element* $z = (z_0, \ldots, z_{n-1})^{\mathrm{t}}$ *is given by the arithmetic mean of its nodes, i.e.,*

$$c(z) := \frac{1}{n} \sum_{k=0}^{n-1} z_k. \tag{5.7}$$

It should be noticed that the centroid of the polygonal element is not necessarily the same as the center of gravity of the domain enclosed by the polygon. However, it serves as a reference point for transformed elements, as is stated by the following result.

Lemma 5.1 (Centroid preservation). *The transformation M given by Definition 5.1 preserves the centroid of the polygon, i.e., $c(z^{(\ell+1)}) = c(z^{(\ell)})$.*

Proof. Using the centroid definition (5.7) and multiplying by n, it has to be shown that

$$\sum_{k=0}^{n-1} z_k^{(\ell+1)} = \sum_{k=0}^{n-1} z_k^{(\ell)},$$

where $z_k^{(\ell+1)}$ denotes the kth node of the transformed polygon $z^{(\ell+1)}$ and $z_k^{(\ell)}$ the kth node of the initial polygon $z^{(\ell)}$. The matrix representation $z^{(\ell+1)} = M z^{(\ell)}$ and rearranging the sums implies

$$\sum_{k=0}^{n-1} z_k^{(\ell+1)} = \sum_{k=0}^{n-1} \left(\sum_{j=0}^{n-1} M_{k,j} z_j^{(\ell)} \right) = \sum_{j=0}^{n-1} \left(\sum_{k=0}^{n-1} M_{k,j} \right) z_j^{(\ell)}.$$

The bracket on the right side represents the sum of the jth column entries of M. Thus, it has to be shown that this equals one. Using the matrix representation (5.4) of M, it follows that

$$\sum_{k=0}^{n-1} M_{k,j} = |w|^2 + |1-w|^2 + w(1-\overline{w}) + \overline{w}(1-w)$$

$$= |w|^2 + (1-w)\overline{(1-w)} + w\overline{(1-w)} + \overline{w}(1-w)$$

$$= |w|^2 + \overline{(1-w)} + \overline{w} - |w|^2 = 1,$$

which implies the result of Lemma 5.1. $\qquad\qquad\square$

Using the same arguments, it follows readily that the two sub transformations given by M' and M'' also preserve the centroid, since the column entries of these matrices also sum up to 1. Due to the similar construction scheme applied to all polygon edges, the transformation matrix M is circulant. Hence, the remarkable theory of circulant matrices can be applied. For the convenience of the reader, the relevant results will be summarized in the following. More detailed descriptions and proofs can be found in [29, 62].

The square matrix $A \in \mathbb{C}^{n \times n}$ is called *circulant*, if it is of the form

$$A := \begin{pmatrix} a_0 & a_1 & \cdots & a_{n-1} \\ a_{n-1} & a_0 & \cdots & a_{n-2} \\ \vdots & \vdots & \ddots & \vdots \\ a_1 & a_2 & \cdots & a_0 \end{pmatrix}.$$

It is fully defined by its first row vector $a := (a_0, \ldots, a_{n-1})$ with zero-based element indices. With $r := \exp(2\pi i/n)$ denoting the nth *root of unity*, it holds that A is diagonalized by the unitary discrete *Fourier matrix*

$$F := \frac{1}{\sqrt{n}} \begin{pmatrix} r^{0\cdot 0} & \cdots & r^{0\cdot(n-1)} \\ \vdots & \ddots & \vdots \\ r^{(n-1)\cdot 0} & \cdots & r^{(n-1)\cdot(n-1)} \end{pmatrix} \qquad (5.8)$$

with entries $F_{j,k} = r^{j \cdot k}/\sqrt{n}$ and likewise zero-based indices $j, k \in \{0, \ldots, n-1\}$. That is, the diagonal matrix D of the eigenvalues η_k, $k \in \{0, \ldots, n-1\}$ of A is given by

$$D = \mathrm{diag}(\eta_0, \ldots, \eta_{n-1}) := F^* A F, \tag{5.9}$$

with F^* denoting the conjugate transpose of F, i.e., $(F^*)_{j,k} = \overline{F_{k,j}}$. Furthermore, the vector $\eta := (\eta_0, \ldots, \eta_{n-1})^t$ of all eigenvalues can easily be computed by multiplying the non-normalized Fourier matrix with the transposed first row of A, i.e., $\eta = \sqrt{n} F a^t$.

One of the remarkable properties of circulant matrices is that they all can be diagonalized using the same basis defined by the columns of the Fourier matrix. With respect to the generalized polygon transformation matrix M, this implies in particular that the basis does not depend on the construction parameters λ and θ. These columns of the Fourier matrix (5.8) itself can be interpreted as polygons, as will be stated by the following definition.

Definition 5.3 (Fourier polygon). *For a natural number $n \geq 3$ and $k \in \{0, \ldots, n-1\}$, the kth column f_k of the Fourier matrix F is denoted as the kth Fourier polygon, i.e.,*

$$f_k := \frac{1}{\sqrt{n}} \left(r^{0 \cdot k}, \ldots, r^{(n-1) \cdot k} \right)^t, \tag{5.10}$$

where $r := \exp(2\pi \mathrm{i}/n)$ is the nth root of unity.

Figure 5.6 depicts all Fourier polygons for $n \in \{3, 4, 5\}$. For each tuple (n, k) the circle with radius $1/\sqrt{n}$ is depicted in gray along with the nodes $r^j/\sqrt{n}$, $j \in \{0, \ldots, n-1\}$. Nodes that are part of a Fourier polygon are marked black. Scaled powers of the nth root of unity that are not part of the polygon are marked gray. To draw the polygon for (n, k) starting from the node representing $z_0 = r^0/\sqrt{n} = 1/\sqrt{n}$, one simply connects each kth node n times, counting the nodes counterclockwise. For the polygons with $k = 0$, depicted in the left column of Figure 5.6, this results in n times the same node z_o. For $k = 1$, the counterclockwise-oriented regular n-gon is obtained, for $k = n - 1$ the clockwise-oriented regular n-gon. The shape of the other cases depend on k being a divisor of n or not. For example, in the case $n = 4$, $k = 2$ a reduced polygon with node multiplicity two occurs.

Comparing the last row of Figure 5.6 with that of Figure 5.5 suggests that the limit figures of the generalized polygon transformation are related to Fourier polygons. This will be analyzed in the following, which starts by deriving the eigenvalues of the generalized polygon transformation.

Lemma 5.2 (Eigenvalues of the generalized transformation). *The eigenvalues of the transformation matrix M given by Definition 5.1 are*

$$\eta_k := \left| 1 - \overline{w} + r^k \overline{w} \right|^2 = |1 - w|^2 + |w|^2 + 2 \operatorname{Re} \left(r^k \overline{w}(1 - w) \right), \tag{5.11}$$

with $k \in \{0, \ldots, n-1\}$.

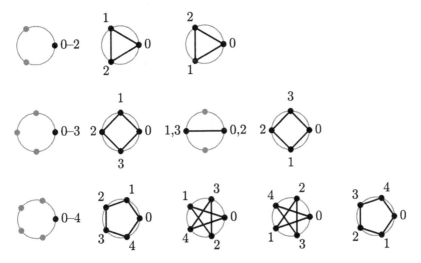

Figure 5.6: Fourier polygons for $n \in \{3, 4, 5\}$ (rows) and $k \in \{0, \ldots, n-1\}$ (columns). Numbers denote node indices in the polygon.

Proof. As stated before, the vector of eigenvalues of an arbitrary circulant matrix A can be computed by multiplying $\sqrt{n}F$ with the transposed first row of A. Hence (5.11) can be obtained by simplifying $\eta_k = \sum_{j=0}^{n-1} r^{k \cdot j} M_{0,j}$, or alternatively, in preparation of following results, by multiplying the eigenvalues

$$\eta'_k = \sum_{j=0}^{n-1} r^{k \cdot j} M'_{0,k} = r^{k \cdot 0}(1 - w) + r^{k \cdot (n-1)} w$$

$$\eta''_k = \sum_{j=0}^{n-1} r^{k \cdot j} M''_{0,k} = r^{k \cdot 0}(1 - \overline{w}) + r^{k \cdot 1} \overline{w}$$

of the circulant matrices M' and M'' having the same eigenvectors. Due to $\overline{r^k} = r^{k(n-1)}$, it follows that $\eta'_k = \overline{\eta''_k}$, which implies that the eigenvalues of M are given by

$$\eta_k = \eta''_k \eta'_k = \eta''_k \overline{\eta''_k} = \left| 1 - \overline{w} + r^k \overline{w} \right|^2.$$

The second representation of η_k given by the right-hand side of (5.11) follows by

$$
\begin{aligned}
|1 - \overline{w} + r^k \overline{w}|^2 &= \left(1 - \overline{w} + r^k \overline{w}\right)\overline{\left(1 - \overline{w} + r^k \overline{w}\right)} \\
&= \left((1 - \overline{w}) + r^k \overline{w}\right)\left((1 - w) + r^{k(n-1)} w\right) \\
&= |1 - w|^2 + r^k \overline{w}(1 - w) + r^{k(n-1)} w(1 - \overline{w}) + r^{kn} |w|^2 \\
&= |1 - w|^2 + |w|^2 + 2 \operatorname{Re}\left(r^k \overline{w}(1 - w)\right)
\end{aligned}
$$

using that $u\overline{u} = |u|^2$, $r^{kn} = 1$, and $u + \overline{u} = 2 \operatorname{Re} u$. $\qquad \square$

From the first representation of η_k in (5.11), it follows readily that all eigenvalues of M are real-valued and positive. Furthermore, it holds that $\eta_0 = 1$ for all $\lambda \in (0,1)$ and $\theta \in (0, \pi/2)$. This is the reason why the generalized polygon transformation has no rotational effect in contrast to the two substeps M' and M'' with complex eigenvalues. This will become apparent from the following decomposition of the transformed polygon

Lemma 5.3 (Transformed polygon decomposition). *The polygon $z^{(\ell)}$ obtained by iteratively applying the generalized polygon transformation ℓ times starting from the initial polygon $z^{(0)}$ is given by*

$$z^{(\ell)} = M^\ell z^{(0)} = \sum_{k=0}^{n-1} \eta_k^\ell v_k, \tag{5.12}$$

where η_k denotes the kth eigenvalue of M according to (5.11) and

$$v_k := \left(F^* z^{(0)}\right)_k f_k = \frac{\left(F^* z^{(0)}\right)_k}{\sqrt{n}} \left(r^{0 \cdot k}, \ldots, r^{(n-1) \cdot k}\right)^{\mathsf{t}} \tag{5.13}$$

denotes the kth decomposition polygon of $z^{(0)}$.

Proof. According to (5.9), $D := F^* M F$ denotes the diagonal matrix with main diagonal entries $D_{k,k} = \eta_k$ representing the eigenvalues of M, and $D_{j,k} = 0$ if $j \neq k$. Thus, the transformation matrix can be written as $M = FDF^*$, since $F^{-1} = F^*$, which implies, for the polygon $z^{(\ell)}$ obtained after applying the transformation ℓ times, $z^{(\ell)} = M^\ell z^{(0)} = (FDF^*)^\ell z^{(0)} = FD^\ell F^* z^{(0)}$. Here, FD^ℓ results in a matrix, where the kth column represents the kth Fourier polygon scaled by η_k^ℓ. This matrix is multiplied by the vector $F^* z^{(0)}$, consisting of the entries $c_k := (F^* z^{(0)})_k$; thus $z^{(\ell)} = \sum_{k=0}^{n-1} \eta_k^\ell c_k f_k = \sum_{k=0}^{n-1} \eta_k^\ell v_k$. □

Due to their definition, the decomposition polygons v_k are scaled and rotated Fourier polygons, since $c_k = \left(F^* z^{(0)}\right)_k$ is complex. Here, $|c_k|$ and the argument of c_k represent the scaling factor and rotation angle of the kth Fourier polygon. In the case $c_k = 0$ the decomposition polygon v_k degenerates to the zero vector. In this case it is impossible to transform the initial polygon $z^{(0)}$ into a polygon similar to the associated Fourier polygon f_k, which is a direct consequence of the representation (5.12).

Samples of initial polygons and their associated decomposition polygons are given in Figure 5.7. Here, the same initial polygons as depicted in the left column of Figure 5.5 have been used. In addition, the first and second nodes are marked by circles and by squares, respectively, to indicate the orientation of each decomposition polygon.

The decomposition polygons are, from left to right, the centroid v_0 with node multiplicity n, the counterclockwise-oriented regular n-gon v_1, and finally the clockwise-oriented regular n-gon v_{n-1}. Between them only star-shaped n-gons occur if n is a prime number. Otherwise, folded polygons with multiple nodes exist, as is also apparent from Figure 5.6 depicting the Fourier polygons. The following lemma summarizes properties of the decomposition polygons.

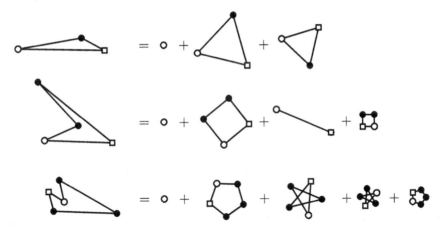

Figure 5.7: Decomposition of polygons. The first two nodes of each polygon are marked by circles and by squares, respectively.

Lemma 5.4 (Decomposition polygon properties). *The decomposition polygons v_k, $k \in \{0, \ldots, n-1\}$, given by (5.13) of a polygon $z^{(0)}$ with $n \geq 3$ nodes, have the following properties:*

1. *The decomposition polygons v_k only depend on the initial polygon $z^{(0)}$, but not on the transformation parameters λ and θ.*

2. *The decomposition polygon v_k is similar to the polygon obtained by successively connecting counterclockwise each kth root of unity starting from the first node $1/\sqrt{n}$.*

3. *The decomposition polygon v_0 is a degenerated polygon representing n times the centroid of $z^{(0)}$.*

4. *The decomposition polygon v_1 represents the counterclockwise-oriented regular n-gon and the decomposition polygon v_{n-1} represents the clockwise-oriented regular n-gon if the associated scaling parameters $\left(F^* z^{(0)}\right)_1$ and $\left(F^* z^{(0)}\right)_{n-1}$ are non-zero.*

5. *The decomposition polygon v_k is a g-fold traversed (n/g)-gon, that is, each node has the multiplicity g, where $g := \gcd(n, k)$ denotes the greatest common divisor of the two natural numbers n and k.*

Proof. Property 1 follows from the definition given by (5.13), only involving the discrete Fourier matrix F and the initial polygon $z^{(0)}$ but not the transformation parameters.

Property 2 is implied by the fact that the decomposition polygons v_k are scaled versions of the Fourier polygons f_k given by (5.10), since the exponents in the nodes $z_j = r^{j \cdot k}/\sqrt{n}$ are all multiples jk of k for $j \in \{0, \ldots, n-1\}$. Hence, the weight $(F^* z^{(0)})_k$ of f_k reflects the scaling and rotation information of the initial polygon $z^{(0)}$ with respect to the associated Fourier polygon.

Property 3 follows directly from

$$\left(F^* z^{(0)}\right)_0 = \sum_{j=0}^{n-1} (F^*)_{0,j} z_j^{(0)} = \frac{1}{\sqrt{n}} \sum_{j=0}^{n-1} z_j^{(0)} \quad \text{and} \quad v_0 = \frac{\left(F^* z^{(0)}\right)_0}{\sqrt{n}} (1, \dots, 1)^{\text{t}}.$$

Property 4 is due to the fact that the powers r^j, $j \in \{0, \dots, n-1\}$ are equidistant and counterclockwise distributed over the unit circle, leading to the regular n-gon v_1. The counterclockwise orientation of v_{n-1} follows from its nodes being scaled versions of $r^{j(n-1)} = r^{jn} r^{-j} = r^{-j}$.

Finally, property 5 again holds due to the scaled Fourier polygon nodes r^{jk}. For $j = g := \gcd(n,k)$ it holds that $r^{jk} = r^n = r^0$. That is, after traversing g nodes, the first node is reached again. This is repeated $n/g - 1$ times leading to a g-fold traversed (n/g)-gon.   □

Property 5 implies that if $\gcd(n,k) = 1$, that is, n is a prime number, all decomposition polygons v_k for $k \in \{2, \dots, n-2\}$ are star shaped n-gons with node multiplicity 1. For the case $n = 5$, this is depicted in the last row of Figure 5.6. Due to the representation $z^{(\ell)} = \sum_{k=0}^{n-1} \eta_k^\ell v_k$ and the fact that the decomposition polygons are fixed, the asymptotic behavior depends on the eigenvalues η_k, which will now be analyzed in detail.

Special cases occur if an eigenvalue becomes zero, since in this case one step of the transformation eliminates the associated decomposition polygon in the decomposition of $z^{(1)} = M z^{(0)}$. To classify these cases, the roots of the eigenvalue functions with respect to the transformation parameters will be determined in the following. Due to symmetry reasons, the case $\lambda = 1/2$ is of special interest. Therefore, this case will be analyzed first by giving an alternative representation of the eigenvalues η_k suitable for deriving intersections and dominance information, that is, information on the largest eigenvalue. Furthermore, the results obtained will also serve in proving the eigenvalue dominance pattern for the general case.

Lemma 5.5 (Eigenvalues of the generalized transformation for $\lambda = 1/2$). *For the special case $\lambda = 1/2$, the eigenvalues η_k of M given by Lemma 5.2 can be written as*

$$\eta_k = \sec^2 \theta \cos^2 \left(\theta - \frac{\pi k}{n} \right), \tag{5.14}$$

where $k \in \{0, \dots, n-1\}$.

Proof. For $\lambda = 1/2$ it holds that $1 - \overline{w} = w$. Together with (5.11) this yields the representation $\eta_k = |w + r^k \overline{w}|^2$. Due to $\theta \in (0, \pi/2)$, it holds that $|w| = \frac{1}{2}\sqrt{1 + \tan^2 \theta} = \frac{1}{2} \sec \theta$. The argument of w is given by θ and that of r^k by $2\pi k/n$. Collecting the real and imaginary parts of the two summands in $w + r^k \overline{w}$ and applying trigonometric identities to the following expressions in squared brackets results in

$$w + r^k \overline{w} = \frac{1}{2} \sec \theta \left(\left[\cos \theta + \cos \left(\frac{2\pi k}{n} - \theta \right) \right] + i \left[\sin \theta + \sin \left(\frac{2\pi k}{n} - \theta \right) \right] \right)$$

$$= \sec \theta \cos \left(\theta - \frac{\pi k}{n} \right) \left(\cos \frac{\pi k}{n} + i \sin \frac{\pi k}{n} \right).$$

Since the absolute value of the complex number in the last bracket is one, this implies the representation (5.14) after squaring. □

By setting $\lambda = 1/2$, the eigenvalues η_k are functions of the base angle $\theta \in (0, \pi/2)$ only. The associated eigenvalue representation (5.14) can now readily be used for deriving the intersections of these functions, which will be given by the following result.

Lemma 5.6 (Eigenvalue intersections for $\lambda = 1/2$). *For $k, m \in \{0, \ldots, n-1\}$, $k \neq m$, the eigenvalue functions η_k and η_m with parameter $\theta \in (0, \pi/2)$ may only intersect for $\theta = j\pi/(2n)$ where $j \in \{1, \ldots, n-1\}$.*

Proof. Since η_k and η_m are positive, this will be shown by analyzing the roots of the difference function

$$d_{k,m} := \eta_k - \eta_m = \sec^2 \theta \left(\cos^2 \left(\theta - \frac{\pi k}{n} \right) - \cos^2 \left(\theta - \frac{\pi m}{n} \right) \right)$$

$$= \sec^2 \theta \sin \left(2\theta - \pi \frac{k+m}{n} \right) \sin \left(\pi \frac{k-m}{n} \right). \tag{5.15}$$

Since $|k - m|/n \in \{1/n, \ldots, (n-1)/n\}$, the last sine factor in (5.15) never becomes zero. Hence, roots can only occur if the argument of the first sine factor is a multiple of π, that is,

$$2\theta - \pi \frac{k+m}{n} = \nu\pi \quad \Leftrightarrow \quad \theta = \frac{\pi}{2} \left(\nu + \frac{k+m}{n} \right), \quad \nu \in \mathbb{Z}.$$

Due to $\theta \in (0, \pi/2)$, the factor $\nu + (k+m)/n$ must be in $(0, 1)$, thus implying $k + m \neq n$ and $\nu = -\lfloor (k+m)/n \rfloor$, where $\lfloor \cdot \rfloor$ denotes the floor function. Since $k \neq m$, this results in $(k + m) \in \{1, \ldots, 2n - 3\} \setminus \{n\}$.

In the case of $(k + m) \in \{1, \ldots, n - 1\}$ and $(k + m) \in \{n + 1, \ldots, 2n - 3\}$, it follows that $\nu = 0$ and $\nu = -1$, respectively, thus providing the roots

$$\theta = \frac{\pi}{2n}(k + m) \quad \text{and} \quad \theta = \frac{\pi}{2n}(k + m - n),$$

respectively, as stated by the lemma. □

For the cases $n \in \{3, 4, 5\}$ and the choice $\lambda = 1/2$, the eigenvalue functions η_k with respect to the base angle $\theta \in (0, \pi/2)$ are depicted in Figure 5.8. Here, eigenvalue function intersections are marked by black circle markers. Intersection points with a change of the dominant eigenvalue are indicated by

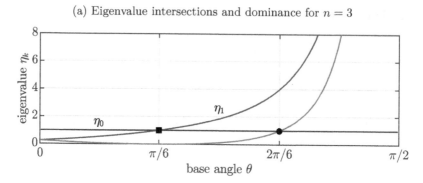

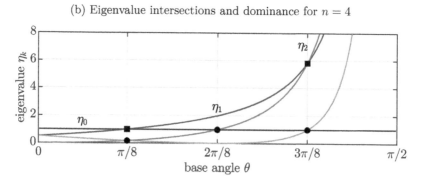

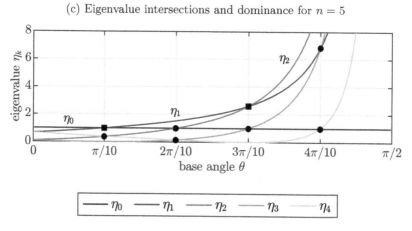

Figure 5.8: Eigenvalue $\eta_k(\theta)$ graph intersections and dominant eigenvalues for $\lambda = 1/2$ and $n \in \{3, 4, 5\}$.

black square markers. In addition, the dominant eigenvalue is noted above the functions. It can be observed that the dominant eigenvalue changes at each odd multiple of $\pi/(2n)$. This is stated by the following lemma.

Lemma 5.7 (Dominant eigenvalue for $\lambda = 1/2$). *For $n \in \mathbb{N}$, $n \geq 3$, $\lfloor \cdot \rfloor$ denoting the floor function, and $k \in \{0, \ldots, \lfloor n/2 \rfloor - 1\}$, characteristic angles are given by*

$$\theta_{-1} := 0, \quad \theta_k := \frac{\pi}{2n}(2k+1), \quad \text{and} \quad \theta_{\lfloor n/2 \rfloor} := \frac{\pi}{2}.$$

On each interval (θ_{k-1}, θ_k), $k \in \{0, \ldots, \lfloor n/2 \rfloor\}$, the index of the dominant eigenvalue of the generalized polygon transformation according to Definition 5.1 for the choice $\lambda = 1/2$ is given by k. That is,

$$\theta \in (\theta_{k-1}, \theta_k) \implies \eta_k > \eta_m \quad \forall m \in \{0, \ldots, n-1\} \setminus \{k\}. \tag{5.16}$$

Furthermore, for $k \in \{0, \ldots, \lfloor n/2 \rfloor - 1\}$,

$$\theta = \theta_k \implies \eta_k = \eta_{k+1} > \eta_m \quad \forall m \in \{0, \ldots, n-1\} \setminus \{k, k+1\}, \tag{5.17}$$

that is, for $\theta = \theta_k$, the eigenvalue function η_k is only intersected by η_{k+1}.

Proof. Induction over the intervals (θ_{k-1}, θ_k) for $k \in \{0, \ldots, \lfloor n/2 \rfloor\}$ will be used. In the case of $k = 0$ implying $\theta \in (\theta_{-1}, \theta_0) = (0, \pi/(2n))$, Equation (5.16) holds, if the associated difference function

$$d_{0,m} = \sec^2 \theta \sin\left(2\theta - \pi \frac{m}{n}\right) \sin\left(\pi \frac{-m}{n}\right)$$

according to (5.15) is positive for all $m \in \{1, \ldots, n-1\}$. This is the case, since the first factor $\sec^2 \theta$ is positive and both sine factors are negative. In the case of the first sine, this is due to $-\pi < 2 \cdot 0 - \pi(n-1)/n < 2\theta - \pi m/n < 2\pi/(2n) - \pi/n = 0$. In the case of the second sine, this follows readily from $-\pi m/n \in (-\pi, 0)$.

To show (5.17), the equation

$$d_{0,m}\left(\frac{\pi}{2n}\right) = \sec^2 \frac{\pi}{2n} \sin\left(\frac{\pi}{n} - \pi \frac{m}{n}\right) \sin\left(\pi \frac{-m}{n}\right) = 0$$

must be solved. Here, the first and the last factor are nonzero. This implies for the argument of the first sine $\pi/n - \pi m/n = \pi(1-m)/n = j\pi$ with $j \in \mathbb{Z}$, hence $m = 1$, since $m \in \{1, \ldots, n-1\}$. That is, for $\theta = \theta_0$ the function η_0 is only intersected by η_1. Furthermore, since all η_k are continuous as well as differentiable on $(0, \pi/2)$ except the points where $\eta_k = 0$, and dominated by η_0 on $(0, \pi/(2n))$, it holds that $1 = \eta_0 = \eta_1 > \eta_m$ for $m \in \{2, \ldots, n-1\}$ in θ_0.

For a given $k \in \{1, \ldots, \lfloor n/2 \rfloor - 1\}$ it will now be assumed that for θ_{k-1} the dominant eigenvalue changes from η_{k-1} to η_k. To prove (5.16) for the interval (θ_{k-1}, θ_k) it suffices to show that η_k is not intersected by any other η_m with $m \in \{0, \ldots, n-1\} \setminus \{k\}$.

According to Lemma 5.6, intersections may only occur for $\theta = \pi k/n$, which is the midpoint of the interval (θ_{k-1}, θ_k). Here $d_{k,m}(\pi k/n) = 0$ implies for the middle factor $\sin(2\pi k/n - \pi(k+m)/n) = 0$, which itself implies $2\pi k/n - \pi(k+m)/n = (\pi/n)(k-m) = j\pi$, $j \in \mathbb{Z}$. Hence it holds that $m = k$ and the dominant function η_k is not intersected by any other η_m inside the interval (θ_{k-1}, θ_k).

Finally, (5.17) is shown by solving the equation

$$d_{k,m}(\theta_k) = \sec^2\theta_k \, \sin\left(\pi\frac{k+1-m}{n}\right) \sin\left(\pi\frac{k-m}{n}\right) = 0.$$

The first sine becomes zero in the case of $m = k + 1$. The second sine, which does not depend on θ, is nonzero due to $m \neq n - k$. Since the difference function changes its sign in θ_k, the dominant eigenvalue changes from η_k on the left of θ_k to η_{k+1} on the right as stated. $\square$

Lemma 5.7 gives a full classification of the eigenvalue intersections and the dominant eigenvalue function for the special case $\lambda = 1/2$. Before a similar result is derived for the general case $\lambda \in (0,1)$, the roots and signs of the eigenvalue functions η_k of the generalized polygon transformation according to Definition 5.1 are derived.

Lemma 5.8 (Roots and signs of the eigenvalues η_k). *For $\theta \in (0, \pi/2)$, $\lambda \in (0,1)$, the eigenvalue η_k is strictly positive, i.e., $\eta_k > 0$, if and only if $k \in \{0, \ldots, \lfloor n/2 \rfloor\}$. Otherwise η_k has exactly one isolated root at $\lambda = 1/2$, $\theta = \pi(2k-n)/(2n)$.*

Proof. Since $\eta_0 = 1$ has no root, it will be assumed that $k \in \{1, \ldots, n-1\}$. According to the representation (5.11) the eigenvalue η_k is positive and its roots are given by the solutions of

$$1 - \overline{w} + r^k\overline{w} = 0 \quad \Leftrightarrow \quad \overline{w} = \frac{1}{1-r^k} = \frac{1}{2}\left(1 + i\cot\frac{\pi k}{n}\right).$$

Using $\overline{w} = \lambda - i(1-\lambda)\tan\theta$ yields $\lambda = 1/2$ and

$$\theta = -\arctan\left(\cot\frac{\pi k}{n}\right) = -\left(\frac{\pi}{2} - \frac{\pi k}{n}\right) = \frac{\pi(2k-n)}{2n}.$$

Due to the restriction $\theta \in (0, \pi/2)$, this implies $k \in \{\lfloor n/2 \rfloor + 1, \ldots, n-1\}$ as stated by the lemma. $\square$

Since $\eta_k' = \overline{\eta_k''}$ and $\eta_k = \eta_k''\eta_k'$, the roots given by Lemma 5.8 also hold for the eigenvalues of M' and M''. In the case of M' and $n = 3$, the choice of transformation parameters $\lambda = 1/2$ and $\theta = \pi/6$ according to Lemma 5.8 yields $\eta_2 = 0$ and $\eta_0 = \eta_1 = 1$. Hence, applying one step of the transformation preserves the centroid of the initial triangle due to $\eta_0 = 1$. Furthermore, due to $\eta_2 = 0$ the decomposition polygon v_2 is eliminated and the decomposition polygon v_1 remains. Since the latter is similar to f_1, representing a counterclockwise-oriented regular triangle, the resulting triangle itself is regular, which proves Napoleon's theorem.

It can also be observed that half of the angles given in Lemma 5.8 are those given by the Petr-Douglas-Neumann theorem. The other half can be obtained by defining the geometric construction of M' not to the right of the segments, but to the left. Hence, the $n-2$ transformations of the Petr-Douglas-Neumann theorem are chosen such that successively $n-2$ of the eigenvalues η_k become zero, thus filtering out the associated decomposition polygons. At the end, the decomposition polygons v_0 and v_1 remain, which implies that the centroid is preserved and the result of the Petr-Douglas-Neumann construction is a regular n-gon.

Based on decomposition polygons, an algebraic foundation for deriving geometric construction schemes transforming arbitrary polygons with n vertices into k-regular n-gons and associated generalized Napoleon vertices are given in [157]. Geometric construction schemes are derived exemplarily for $n \in \{3,4,5\}$, $k = 1$ and $n = 5$, $k = 2$. Naturally, such schemes only exist if the associated polygon is constructible. For the regular n-gon, Gauß showed in [58] that this is the case if n is a product of a power of two and any number of distinct Fermat prime numbers $F_m = 2^{(2^m)} + 1$. Furthermore, fractal polygons have been derived in [161] by combining prime indexed Fourier polygons and an approximate prime counting function.

In the following, the behavior of the polygon $z^{(\ell)}$ will be analyzed with respect to the parameters λ and θ if ℓ tends to infinity. According to the decomposition (5.12), this depends on the dominant eigenvalue η_k, i.e., the eigenvalue with the largest magnitude. To motivate upcoming results, Figure 5.9 depicts top-down visualizations of the eigenvalues η_k for $k \in \{3,4,5\}$.

The meshes on the left side show the surfaces obtained by plotting the bivariate eigenvalue functions η_k with respect to their parameters $\theta \in (0, \pi/2)$ and $\lambda \in (0,1)$. Results for the eigenvalue $\eta_0 = 1$, representing the centroid, are depicted dark gray. Those of η_1, representing the regular n-gon, are medium gray. Subsequent eigenvalues are marked by different shades of light gray. Here, the same shades of gray have been used as in Figure 5.8, showing the intersections of these surfaces with the plane $\lambda = 1/2$.

Since the eigenvalue with the largest magnitude is of particular interest, the right side shows the partition of the parameter domain shaded according to the dominant eigenvalue. That is, the right side shows an orthogonal view from above on the surfaces depicted on the left side. For example, the point (θ, λ) is shaded dark gray if $\eta_1 > \eta_k$ for all $k \in \{0, 2, \ldots, n-1\}$. It is shaded almost black if $\eta_0 = 1 > \eta_k$ for all $k \in \{1, \ldots, n-1\}$. From a mesh smoothing point of view, the dominance of η_1 corresponding to the counterclockwise-oriented regular n-gon is of particular interest. Thus, only a fixed parameter pair (θ, λ) in the dark gray subdomains depicted on the right side of Figure 5.9 should be used for transformation.

In generalization of the eigenvalue intersections for the special case $\lambda = 1/2$ derived in Lemma 5.6, the eigenvalue intersections for $\lambda \in (0,1)$ must be

(a) Eigenvalue intersections and dominance for $n = 3$

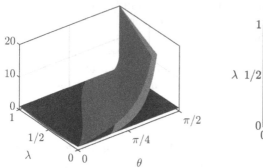

(b) Eigenvalue intersections and dominance for $n = 4$

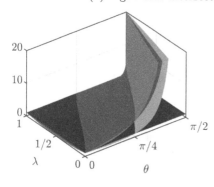
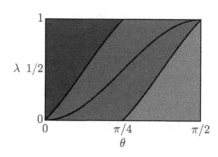

(c) Eigenvalue intersections and dominance for $n = 5$

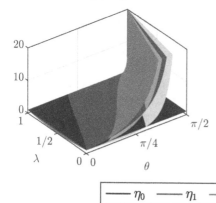
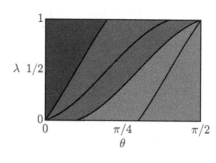

Figure 5.9: Eigenvalue $\eta_k(\theta, \lambda)$ graph intersections for $\theta \in (0, \pi/2)$, $\lambda \in (0, 1)$ (left) and dominant eigenvalue domains (right) for $n \in \{3, 4, 5\}$.

derived first to give a proper definition of the suitable parameter domains for defining regularizing polygon transformations.

Lemma 5.9 (Eigenvalue intersections). *For $\theta \in (0, \pi/2)$, $\lambda \in (0, 1)$, the eigenvalues η_k, η_m with $k, m \in \{0, \ldots, n-1\}$ and $k \neq m$ intersect only if $(k+m) \bmod n \neq 0$. In that case $\eta_k = \eta_m$ only holds along the parameter line*

$$\lambda_{k,m}(\theta) := \sin\theta \left(\sin\theta + \cot\left(\frac{\pi}{n}(k+m)\right) \cos\theta \right), \tag{5.18}$$

where $\theta \in (0, \pi/2)$ and $0 < \lambda_{k,m}(\theta) < 1$. Furthermore, the eigenvalues only intersect pairwise.

Proof. According to (5.11) the eigenvalues are positive real-valued. Therefore, eigenvalue intersection parameters can be obtained by finding the roots of the difference function $\eta_k - \eta_m$, hence the solutions of

$$\text{Re}\left((r^k - r^m)\overline{w}(1-w) \right) = \text{Re}\left((r^k - r^m)(\overline{w} - |w|^2) \right) = 0. \tag{5.19}$$

Using the complex representation $r^k = \sin(2\pi k/n) + i\cos(2\pi k/n)$ and angle addition theorems results in the representation

$$r^k - r^m = -2\sin\left(\frac{\pi}{n}(k-m)\right) \underbrace{\left(\sin\left(\frac{\pi}{n}(k+m)\right) - i\cos\left(\frac{\pi}{n}(k+m)\right) \right)}_{=:u_1}.$$

For the second factor of the argument on the right side of (5.19), it holds that

$$\overline{w} - |w|^2 = (1-\lambda) \underbrace{\left((\lambda - (1-\lambda)\tan^2\theta) - i\tan\theta \right)}_{=:u_2}.$$

Hence equation (5.19) simplifies to

$$-2\sin\left(\frac{\pi}{n}(k-m)\right)(1-\lambda)\,\text{Re}(u_1 u_2) = 0.$$

Since $(k-m) \in \{-(n-1), \ldots, (n-1)\} \setminus \{0\}$ and $\lambda \in (0,1)$, the factors preceding $\text{Re}(u_1 u_2)$ are nonzero. Therefore, the solutions are characterized by $\text{Re}(u_1 u_2) = 0$, which is

$$\sin\left(\frac{\pi}{n}(k+m)\right)\left(\lambda - (1-\lambda)\tan^2\theta\right) - \cos\left(\frac{\pi}{n}(k+m)\right)\tan\theta = 0.$$

In the case of $(k+m) \bmod n = 0$, this equation reduces to $\tan\theta = 0$, which has no solution for $\theta \in (0, \pi/2)$. Otherwise, rearranging the terms leads to the equation

$$\lambda(1 + \tan^2\theta) - \tan^2\theta = \cot\left(\frac{\pi}{n}(k+m)\right)\tan\theta.$$

Since the left side is linear in λ, the following explicit representation of the intersection line of the eigenvalues η_k and η_m can be derived

$$\lambda = \frac{1}{1 + \tan^2 \theta} \tan \theta \left(\tan \theta + \cot \left(\frac{\pi}{n}(k+m) \right) \right),$$

which simplifies to (5.18).

The eigenvalues intersect only pairwise since there is no index triple $k, m, j \in \{0, \ldots, n-1\}$ with $k < m < j$ fulfilling $(k+m) \bmod n = j$ and $(k+j) \bmod n = j$ where $j \in \{1, \ldots, n-1\}$.   □

Due to the sum of indices in the representation (5.18) and the periodicity of the cotangent function, all index pairs $k, m \in \{0, \ldots, n-1\}$ with $(k+m) \bmod n = j$ share the same line of intersection parameters, which can be written as $\lambda_{0,j}(\theta)$. Furthermore, since $k \neq m$, $k + m \neq n$, there are only $n-1$ distinct parameter lines with $j \in \{1, \ldots, n-1\}$.

The intersection lines $\lambda_{0,j}$ are depicted on the right side of Figure 5.9. The representation (5.18) and the monotonicity of the cotangent function imply that the intersection lines do not intersect each other. Furthermore, it holds that $\lim_{\theta \to 0} \lambda_{0,j}(\theta) = 0$ and $\lim_{\theta \to \pi/2} \lambda_{0,j}(\theta) = 1$. Hence, the parameter lines (5.18) lead to a natural partition of the parameter domain. In Figure 5.9, the resulting subdomains are shaded according to the dominant eigenvalues. As can be seen, there is a change at each second intersection line, which leads to the following partition.

Definition 5.4 (Dominant eigenvalue parameter domains). *A partition of the parameter domain*

$$D := \{(\theta, \lambda) \mid \theta \in (0, \pi/2), \ \lambda \in (0, 1)\}$$

into subdomains D_k and eigenvalue intersection sets S_k is given by

$$D = \left(\bigcup_{k=0}^{\lfloor n/2 \rfloor} D_k \right) \cup \left(\bigcup_{k=0}^{\lfloor n/2 \rfloor - 1} S_k \right), \tag{5.20}$$

with

$$D_k := \{(\theta, \lambda) \mid \theta \in (0, \pi/2), \ \underline{\lambda}_k(\theta) < \lambda < \overline{\lambda}_k(\theta)\}, \tag{5.21}$$
$$S_k := \{(\theta, \lambda) \mid \theta \in (0, \pi/2), \ \lambda = \lambda_{k,k+1}(\theta)\} \cap D, \tag{5.22}$$

where

$$\underline{\lambda}_k(\theta) := \begin{cases} 0 & \text{if } k = \lfloor n/2 \rfloor \\ \max\left(0, \lambda_{0,2k+1}(\theta)\right) & \text{otherwise} \end{cases}$$

and

$$\overline{\lambda}_k(\theta) := \begin{cases} 1 & \text{if } k = 0 \\ \min\left(1, \lambda_{0,2k-1}(\theta)\right) & \text{otherwise.} \end{cases}$$

Here, $\lfloor \cdot \rfloor$ denotes the floor function.

The following lemma substantiates the assumption on the pattern in which the dominant eigenvalue changes, which resulted in the partition of the parameter domain according to Definition 5.4.

Lemma 5.10 (Dominant eigenvalue). *For the parameter choice $(\theta, \lambda) \in D_k$, $k \in \{0, \ldots, \lfloor n/2 \rfloor\}$, it holds that $\eta_k > \eta_m$ for all $m \in \{0, \ldots, n-1\} \setminus \{k\}$. In the case of $(\theta, \lambda) \in S_k$, $k \in \{0, \ldots, \lfloor n/2 \rfloor - 1\}$, it holds that $\eta_k = \eta_{k+1} > \eta_m$ for all $m \in \{0, \ldots, n-1\} \setminus \{k, k+1\}$.*

Proof. In Lemma 5.7 this has already been shown for the restriction to $\lambda = 1/2$. I.e., it has been shown for the eigenvalues η_k of the generalized transformation and parameters in $\widehat{D} := \{(\theta, \lambda) \mid \theta \in (0, \pi/2), \lambda = 1/2\}$ that for $(\theta, \lambda) \in \widehat{D} \cap D_k$, $k \in \{0, \ldots, \lfloor n/2 \rfloor\}$, it holds that $\eta_k > \eta_m$ for all $m \in \{0, \ldots, n-1\} \setminus \{k\}$. Since the eigenvalue functions η_k are continuous on D_k and according to Lemma 5.9 do not intersect each other within D_k, this dominance pattern also holds for η_k if $(\theta, \lambda) \in D_k$ as stated by the lemma.

According to Lemma 5.9 the eigenvalues η_k and η_{k+1} intersect for parameters $(\theta, \lambda) \in S_k$, $k \in \{0, \ldots, \lfloor n/2 \rfloor - 1\}$. Since these eigenvalues are continuous on D and dominant in the neighboring domains D_k and D_{k+1}, they are also dominant in S_k. Furthermore, since eigenvalues intersect only pairwise, there is no other η_m, $m \in \{0, \ldots, n-1\} \setminus \{k, k+1\}$ with $\eta_m = \eta_{k+1} = \eta_m$, which proves the second part of Lemma 5.10. $\square$

Due to the decomposition (5.12), the dominant eigenvalue determines the shape of the polygon $z^{(\ell)} = M^\ell z^{(0)}$ if ℓ tends to infinity. This is depicted in Figure 5.10, which shows the resulting polygons for selected iteration steps ℓ. Here, the pentagon depicted on the lower left of Figure 5.7 has been used as an initial polygon. For each domain D_k, or set S_k in the decomposition (5.20) of D, one parameter pair (θ, λ) has been chosen. All polygons have been scaled with respect to their centroids by $(1/\eta_{\max})^\ell$, where $\eta_{\max} := \max_{k \in \{0, \ldots, n-1\}} \eta_k$ denotes the maximal eigenvalue.

As can be seen, for $(\theta, \lambda) \in D_0$ the polygon degenerates to its centroid, whereas in the case of $(\theta, \lambda) \in D_1$ it becomes a counterclockwise-oriented regular pentagon. These limit figures, as well as the star-shaped pentagon resulting from $(\theta, \lambda) \in D_2$, can also be found in the decomposition of $z^{(0)}$ according to Figure 5.7. For $(\theta, \lambda) \in S_k$, $k \in \{0, \ldots, \lfloor n/2 \rfloor - 1\}$, the resulting limit polygons are linear combinations of the neighboring decomposition polygons, as will be stated by the following lemma.

Lemma 5.11 (Limit polygons). *For $\ell \in \mathbb{N}_0$ let*

$$z_s^{(\ell)} := v_0 + \frac{1}{\eta_{\max}^\ell}(z^{(\ell)} - v_0) = v_0 + \sum_{k=1}^{n-1} \left(\frac{\eta_k}{\eta_{\max}}\right)^\ell v_k \qquad (5.23)$$

denote the polygon $z^{(\ell)} = M^\ell z^{(0)}$ scaled with respect to the centroid v_0 by the

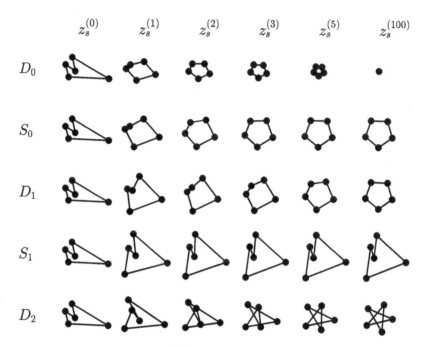

Figure 5.10: Scaled polygons $z_s^{(\ell)}$ for different iteration numbers ℓ and construction parameters (θ, λ) chosen within the given parameter set.

inverse of the ℓth power of the maximal eigenvalue. Then $z^{(\ell)}$ tends to

$$z_s^{(\infty)} := \begin{cases} v_0 & \text{if } (\theta, \lambda) \in D_0 \\ v_0 + v_k & \text{if } (\theta, \lambda) \in D_k, k \in \{1, \ldots, \lfloor n/2 \rfloor\} \\ v_0 + v_1 & \text{if } (\theta, \lambda) \in S_0 \\ v_0 + v_k + v_{k+1} & \text{if } (\theta, \lambda) \in S_k, k \in \{1, \ldots, \lfloor n/2 \rfloor - 1\}, \end{cases} \quad (5.24)$$

that is, $z_s^{(\infty)} = \lim_{\ell \to \infty} z_s^{(\ell)}$.

Proof. The representation (5.24) will be shown by analyzing the eigenvalue quotients $\varrho_k := \eta_k / \eta_{\max} \in [0, 1]$, $k \in \{1, \ldots, n-1\}$, in the sum on the right-hand side of (5.23), which is implied by (5.12) and $\eta_0 = 1$. In the case of $\eta_{\max} = \eta_k$, it holds that $\varrho_k^\ell = 1$. Hence, the associated decomposition polygon v_k is kept unscaled in the sum. Otherwise $\varrho_k < 1$ implies that $\varrho_k^\ell v_k$ tends to the zero vector if ℓ tends to infinity. Hence, the limit of the sum on the right-hand side of (5.23) is given by the sum of the decomposition polygons belonging to the maximal eigenvalues. The latter are given by Lemma 5.10.   □

It should be noticed that in the case $k = 0$ the maximal eigenvalue is given by $\eta_0 = 1$. Hence for $(\theta, \lambda) \in D_0 \cup S_0$ it holds that $z_s^{(\ell)} = z^{(\ell)}$. The unscaled sequence of polygons $z^{(\ell)}$ degenerates for parameter pairs in D_0, converges to a bounded polygon in S_0, and diverges otherwise. Nevertheless, the transformation has a regularizing effect, since the behavior depends on the dominant term.

Lemma 5.11 also demonstrates the advantage of combining the basic transformations M' and M''. Since all eigenvalues of $M = M''M'$ are positive, each step of the transformation has only a scaling effect with respect to the decomposition polygons and their associated eigenvalues. Hence scaling with $(1/\eta_{\max})^{\ell}$ leads to a converging sequence of polygons $z_s^{(\ell)}$. In contrast, the eigenvalues of M' and M'', respectively, are complex-valued, which implies that one step of the transformation not only scales each decomposition polygon according to the absolute value of its associated eigenvalue, but also rotates it by an angle given by the argument of the eigenvalue. Nevertheless, as has been shown in [155] for the special case $\lambda = 1/2$, there exist $2n$ converging subsequences where the limit polygons for odd and even ℓ differ only in a cyclic shift of indices, but not in geometry.

Theorem 5.1 (Regularizing polygon transformation). *For each specific choice of construction parameters $(\theta, \lambda) \in D_1$ according to (5.21) the resulting transformation matrix M according to (5.4) is a regularizing polygon transformation. I.e., for an arbitrary initial polygon $z^{(0)} \in \mathbb{C}^n$ with the inner product $\langle z^{(0)}, f_1 \rangle \neq 0$, it follows that $z^{(\ell)} = M^{\ell} z^{(0)}$ becomes similar to a regular counterclockwise-oriented n-gon if ℓ tends to infinity. Furthermore, the centroid of the transformed polygon is preserved and transforming a regular counterclockwise-oriented n-gon has only a scaling but not a rotational effect.*

Proof. For transformation parameters $(\theta, \lambda) \in D_1$, with D_1 given by (5.21), Lemma 5.11 implies that $z^{(\ell)}$ becomes similar to $v_0 + v_1$ if ℓ tends to infinity. That is, $z^{(\ell)}$ becomes a regular, counterclockwise-oriented polygon if v_1 is not the zero vector. According to (5.13) it holds that $v_1 = (F^* z^{(0)})_1 f_1$, which is not the zero vector if $(F^* z^{(0)})_1 \neq 0$. This implies

$$(F^* z^{(0)})_1 = \sum_{k=0}^{n-1} (F^*)_{1,k} z_k^{(0)} = \sum_{k=0}^{n-1} \overline{F_{k,1}} z_k^{(0)} = \sum_{k=0}^{n-1} \overline{\sqrt{n}(f_1)_k} z_k^{(0)}$$

$$= \sqrt{n} \sum_{k=0}^{n-1} \overline{(f_1)_k} z_k^{(0)} = \sqrt{n} \langle z^{(0)}, f_1 \rangle \neq 0$$

as stated by the theorem using that $\langle z, w \rangle = \sum_{k=0}^{n-1} z_k \overline{w_k}$. The preservation of the centroid has been shown by Lemma 5.1 and the scaling but not rotational effect of M is due to $\eta_k \in \mathbb{R}_0^+$ for all $k \in \{0, \ldots, n-1\}$, which follows from the representation (5.11). $\square$

5.1.2.3 Practical aspects of the regularizing polygon transformation

The regularizing polygon transformation is given in Definition 5.1 as complex circulant matrix to ease the analysis of transformation properties. However, in practice the element transformation is usually not implemented as sparse matrix multiplication. Instead, using (5.4), the new iterate z_k'' of the node z_k being transformed by M is represented as a complex linear combination of the node itself and its predecessor z_{k-1} and successor z_{k+1} according to

$$z_k'' = w(1 - \overline{w})z_{k-1} + \left(|w|^2 + |1 - w|^2\right) z_k + \overline{w}(1 - w)z_{k+1}, \qquad (5.25)$$

where $k \in \{0, \ldots, n - 1\}$. Again, all indices must be taken modulo n. By using the representations $z_k = x_k + iy_k$ and $w = \lambda + i(1 - \lambda)\tan\theta$ one can derive a real-valued representation of the transformation in a vectorial sense by expanding the complex multiplications and collecting the real and imaginary parts.

Lemma 5.12 (Real-valued regularizing polygon transformation). *For the three successive polygon nodes $p_{k-1}, p_k, p_{k+1} \in \mathbb{R}^2$ and the transformation parameters $\lambda \in (0, 1)$ and $\theta \in (0, \pi/2)$, the new node p_k'' obtained by the regularizing polygon transformation is given by*

$$p_k'' = c_1 \begin{pmatrix} y_{k+1} - y_{k-1} \\ x_{k-1} - x_{k+1} \end{pmatrix} + c_2 \begin{pmatrix} x_{k-1} + x_{k+1} \\ y_{k-1} + y_{k+1} \end{pmatrix} + c_3 \begin{pmatrix} x_k \\ y_k \end{pmatrix} \qquad (5.26)$$

with the real-valued constants

$$c_1 := (1 - \lambda)\tan\theta, \quad c_2 := \lambda(1 - \lambda) - c_1^2, \quad c_3 := 1 - 2c_2.$$

Proof. According to (5.25), z_k'' is a weighted linear combination of the nodes z_{k-1}, z_k, and z_{k+1}. For the weights it holds that $w = \lambda + i(1-\lambda)\tan\theta = \lambda + ic_1$, which implies

$$w(1 - \overline{w}) = (\lambda + ic_1)\left((1 - \lambda) + ic_1\right) = \lambda(1 - \lambda) - c_1^2 + i\left(c_1(1 - \lambda) + c_1\lambda\right)$$
$$= \left(\lambda(1 - \lambda) - c_1^2\right) + ic_1 = c_2 + ic_1$$

and thus $\overline{w}(1 - w) = \overline{w(1 - \overline{w})} = c_2 - ic_1$. Furthermore

$$|w|^2 + |1 - w|^2 = |\lambda + ic_1|^2 + |(1 - \lambda) - ic_1|^2 = \lambda^2 + c_1^2 + (1 - \lambda)^2 + c_1^2$$
$$= 1 - 2\lambda(1 - \lambda) + 2c_1^2 = 1 - 2c_2 = c_3.$$

Using the complex representation $z_k = x_k + iy_k$ and similar representations for z_{k-1} and z_{k+1}, Equation (5.25) can be written as

$$z_k'' = (c_2 + ic_1)(x_{k-1} + iy_{k-1}) + c_3(x_k + iy_k) + (c_2 - ic_1)(x_{k+1} + iy_{k+1})$$
$$= (c_2 x_{k-1} - c_1 y_{k-1} + c_3 x_k + c_2 x_{k+1} + c_1 y_{k+1})$$
$$\quad + i(c_1 x_{k-1} + c_2 y_{k-1} + c_3 y_k - c_1 x_{k+1} + c_2 y_{k+1}).$$

Using that $(p_k'')_0 = \text{real}(z_k'')$ and $(p_k'')_1 = \text{imag}(z_k'')$ results in (5.26). $\square$

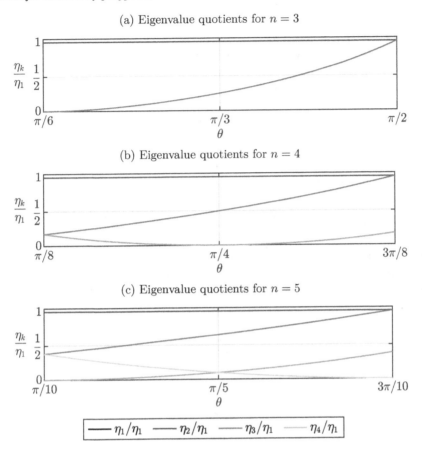

Figure 5.11: Eigenvalue quotients η_k/η_1 indicating the rate of regularization for $\lambda = 1/2$ and $\theta \in [\pi/(2n), 3\pi/(2n)]$ for $n \in \{3, 4, 5\}$.

Due to its construction, iteratively transforming a polygon with transformation parameters taken from D_1 leads to a more regular but also larger element. According to Lemma 5.11, scaling by the inverse of the maximal eigenvalue is a natural choice to prevent element growth. Since η_1 is the dominant eigenvalue in D_1 it follows from (5.23) that $z_s^{(\ell)} = v_0 + v_1 + \sum_{k=2}^{n-1} (\eta_k/\eta_1)^\ell v_k$. Thus, the ratios η_k/η_1 give an estimate of the reduction rate of each decomposition polygon v_k per iteration. These ratios are depicted in Figure 5.11 for $n \in \{3, 4, 5\}$ and the special choice $\lambda = 1/2$. Together with $(\theta, \lambda) \in D_1$ this implies $\theta \in (\pi/(2n), 3\pi/(2n))$.

As can be seen for this choices of θ, the eigenvalue η_1 associated to the regular counterclockwise-oriented polygon v_1 is the dominant eigenvalue. Focusing on the case $n = 4$ depicted in the middle and the special choice $\theta = \pi/4$, one can see that applying one step of the transformation and scaling the resulting polygon by $1/\eta_1$ preserves the decomposition polygon v_1 due to the ratio

$\eta_1/\eta_1 = 1$. Furthermore, v_2 is scaled by the factor $1/2$ due to $\eta_2/\eta_1 = 1/2$ and v_3 is eliminated due to $\eta_3/\eta_1 = 0$.

An even faster convergence rate with respect to v_2 can be achieved by choosing θ slightly larger than $\pi/8$, which also leads to a gradual reduction of v_3. In contrast, if θ tends to $3\pi/8$, the ratio η_2/η_1 tends to one, resulting in a slow regularization, since v_2 is only moderately damped. Thus, the rate of regularization can be controlled by an appropriate choice of transformation parameters.

The scaling factor $1/\eta_1$ is a suitable choice from an asymptotic point of view. However, in this case the size of the limit polygon depends on the size of the decomposition polygon v_1, which may differ significantly from the size of the initial polygon. Thus, in practice, different scaling schemes are applied, which are more related to the actual size of the polygon. A natural scaling variant is given by the following definition.

Definition 5.5 (Edge length scaling). *Let $p_0, \ldots, p_{n-1} \in \mathbb{R}^2$ be the nodes of a planar polygon $P = (p_0, \ldots, p_{n-1})$ with centroid $c := \frac{1}{n}\sum_{k=0}^{n-1} p_k$. The edge length of the polygon is given by*

$$\mathrm{el}(P) := \sum_{k=0}^{n-1}\|p_{k+1} - p_k\|$$

with indices taken modulo n.

The nodes $\hat{p}_k''$ of the edge length scaled transformed polygon $\hat{P}''$ are given by

$$\hat{p}_k'' := c + \frac{\mathrm{el}(P)}{\mathrm{el}(P'')}\,(p_k'' - c),$$

where p_k'' are the nodes of the polygon P'' obtained by applying one step of the regularizing polygon transformation according to (5.26) to P.

The centroid c of the polygon is a natural scaling center, since it is preserved by the transformation. Thus, c is also the centroid of $\hat{P}''$. The edge length scaling scheme given by Definition 5.5 is used to compensate the change in polygon size due to the transformation. As described before, the rate of regularization can be controlled by the appropriate choice of the transformation parameters $(\theta, \lambda) \in D_1$. However, as is depicted in Figure 5.11, some of the ratios η_k/η_1 can become very small, resulting in a significant change of shape. In addition, due to (5.23), applying the iteration ℓ times results in an exponential damping of the decomposition polygon v_k by $(\eta_k/\eta_1)^\ell$. From a smoothing point of view, an additional mechanism to compensate these effects is desirable. It is provided by the following relaxation mechanism based on using a linear combination of the old and the new transformed polygon. Here, the edge length scaled transformed polygon is used to blend between two polygons of the same size.

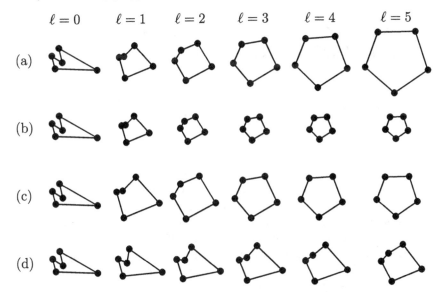

$$\ell = 0 \qquad \ell = 1 \qquad \ell = 2 \qquad \ell = 3 \qquad \ell = 4 \qquad \ell = 5$$

(a)

(b)

(c)

(d)

Figure 5.12: Sequences obtained by iterative transformation using $\theta = 3\pi/20$ and $\lambda = 1/2$. (a) original sequence $P^{(\ell)}$, (b) scaled by dominant eigenvalue $1/\eta_1 \approx 0.8138$, (c) edge length scaled, and (d) edge length scaled and relaxed using $\varrho = 1/3$.

Definition 5.6 (Polygon relaxation). *Let P denote a polygon and $\hat{P}''$ the associated edge length scaled transformed polygon according to Definition 5.5. The nodes $\tilde{p}''_k$ of the relaxed edge length scaled transformed polygon $\hat{P}''$ are given by*

$$\tilde{p}''_k := (1 - \varrho)p_k + \varrho\hat{p}''_k \tag{5.27}$$

using a relaxation factor $\varrho \in [0, 1]$ and $k \in \{0, \ldots, n - 1\}$.

The parameter ϱ blends between the original polygon P represented by $\varrho = 0$ and its edge length scaled counterpart $\hat{P}''$ represented by $\varrho = 1$. The smaller ϱ gets, the smaller is the effect of the transformation. Again, the centroid is preserved by the relaxation operation. Thus, the relaxed edge length scaled transformation scheme given by (5.27) provides all required parameters to control the transformation with respect to various aspects. The regularizing effect and the rate of convergence is controllable by the transformation parameters (θ, λ) and the relaxation parameter ϱ. The preservation of size is ensured by the edge length-based scaling scheme. Effects of these control mechanisms are depicted in Figure 5.12.

The first row denoted by (a) shows the sequence obtained by iteratively applying the regularizing polygon transformation for the choice $\theta = 3\pi/20$ and $\lambda = 1/2$. According to (5.11), in this case it holds that $\eta_1 \approx 1.2288$. Thus, the polygons of the sequence $P^{(\ell)}$ grow rather moderately. Nevertheless, such

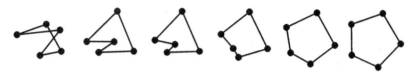

Figure 5.13: Untangling of a random pentagon by the generalized transformation using $\lambda = 1/2$ and $\theta = \pi/4$. Depicted from left to right are the iterations 0, 1, 2, 5, 10, and 15.

an element growth is undesirable in a mesh smoothing context. It can be prevented by scaling each polygon after transformation by the natural scaling factor $1/\eta_1 \approx 0.8138$. This results in the sequence $P_s^{(\ell)}$ as defined in (5.23), which is depicted in row (b). Here, the sequence tends to the decomposition polygon v_1. Again, the size of v_1 might also differ significantly from that of the initial polygon $P^{(0)}$. Thus, applying edge length-based scaling according to Definition 5.5 instead yields the sequence $\hat{P}^{(\ell)}$ shown in row (c), which is more stable with respect to element size.

For a fixed iteration step $\ell \in \mathbb{N}_0$, the polygons of the rows (a) to (c) only differ in size but not in shape or location. All sequences have in common that five iteration steps suffice to result in an almost regular pentagon. This is due to small ratios $\eta_k/\eta_1 < 0.5125$ for $k \in \{2, 3, 4\}$ for the given choice of transformation parameters. If a more moderate rate of convergence is desired, polygon relaxation according to Definition 5.6 can be applied after each transformation and scaling step. This is depicted in row (d) for $\varrho = 1/3$. As can be seen, this reduces the rate of regularization such that the polygon obtained after five transformation steps is more similar to those after two transformation steps of the edge length scaled sequence depicted in (c). Thus, in a mesh smoothing context, edge length-based scaling and the relaxation parameter ϱ are effective concepts to control the sequence of transformed polygons with respect to both element size and rate of convergence.

As has been shown previously, the regularizing polygon transformation iteratively damps all decomposition polygons except the centroid and the regular counterclockwise decomposition polygon of a polygon. Due to this general approach, the regularizing transformation is also suitable for untangling polygonal elements as is depicted in Figure 5.13 for a random initial pentagon with self-intersections.

It has been successively transformed using $\lambda = 1/2$ and $\theta = \pi/4$. In addition, edge length scaling has been applied after each transformation step. As can be seen, the self-intersection is already removed by the first transformation step. However, since the transformation acts as a filter for the regular counterclockwise decomposition polygon, it fails in situations where this decomposition polygon is missing. The following lemma gives a characterization of such polygons.

Lemma 5.13 (Unregularizable polygons). *A planar polygon P with $n \geq 3$ nodes $p_k := (x_k, y_k)^t \in \mathbb{R}^2$, $k \in \{0, \ldots, n-1\}$, satisfying the equations*

$$\sum_{k=0}^{n-1} \left(\cos\left(\frac{2\pi k}{n}\right) x_k + \sin\left(\frac{2\pi k}{n}\right) y_k \right) = 0 \qquad and$$

$$\sum_{k=0}^{n-1} \left(\cos\left(\frac{2\pi k}{n}\right) y_k - \sin\left(\frac{2\pi k}{n}\right) x_k \right) = 0,$$

cannot be regularized by the polygon transformation given by Lemma 5.12.

Proof. According to the proof of Theorem 5.1 a polygon $z^{(0)}$ with complex-valued nodes $z_k = x_k + iy_k$ is not regularizable, if $\langle z^{(0)}, f_1 \rangle = 0$, where $f_1 = (1/\sqrt{n})(r^0, \ldots, r^{(n-1)})^t$ denotes the second Fourier polygon and $r = \exp(2\pi i/n)$ the first root of unity. Using that

$$\sqrt{n}(f_1)_k = r^k = \exp(2\pi k i/n) = \cos\left(\frac{2\pi k}{n}\right) + i \sin\left(\frac{2\pi k}{n}\right)$$

and $\langle z^{(0)}, f_1 \rangle = \sum_{k=0}^{n-1} z_k \overline{(f_1)_k}$ implies

$$\langle z^{(0)}, f_1 \rangle = \frac{1}{\sqrt{n}} \sum_{k=0}^{n-1} (x_k + iy_k) \left(\cos\left(\frac{2\pi k}{n}\right) - i \sin\left(\frac{2\pi k}{n}\right) \right).$$

Expanding the complex products and collecting the summands of the real and imaginary parts of $\langle f_1, z^{(0)} \rangle$ yields the equations as stated by the lemma. $\square$

Lemma 5.13 can be used to identify problematic polygons. Such polygons can be easily regularized if a scaled version of f_1 is added. That is, by using the new node coordinates

$$\tilde{x}_k = x_k + \sigma \cos\left(\frac{2\pi k}{n}\right) \quad and \quad \tilde{y}_k = y_k + \sigma \sin\left(\frac{2\pi k}{n}\right),$$

with an arbitrarily chosen scaling factor $\sigma > 0$. Alternatively, and less systematically, all polygon nodes can be randomly distorted. In practice, such modifications are not required, since mesh elements often fulfill other conditions implying their regularizability.

It is emphasized again that the regularizing polygon transformation was chosen such that the resulting limit polygon is counterclockwise-oriented. Clockwise limit polygons can be obtained by constructing the similar polygons on the left side of the edges instead of on the right side. Hence, the transformation must be defined taking the desired orientation into account.

The robustness of the regularizing transformation is demonstrated by the following numerical test for triangles, quadrilaterals, and pentagons, i.e., $n \in \{3, 4, 5\}$. For each type, 100,000 initial polygons P_j with random node coordinates have been generated. Furthermore, 101 values θ_k have been chosen, evenly spacing the interval $[\pi/(2n), 3\pi/(2n)]$. For each pair (P_j, θ_k) taken from the Cartesian product of all polygons of one type and all θ_k values, the generalized regularizing polygon transformation using the parameters $\lambda = 1/2$ and θ_k has been iteratively applied, starting with the initial polygon P_j until the mean ratio quality number of the resulting polygon has deviated less than 10^{-6} from the ideal value one. After each iteration step the resulting polygon has been scaled to preserve its initial edge length according to Definition 5.5 to avoid numerical instabilities caused by the successive element growth. For each polygon type this test resulted in 10,100,000 transformation cycles, each represented by its number of iterations, required in order to result in an almost regular polygon.

Results of this test are depicted in Figure 5.14. Here, for each polygon type and specific choice of θ_k, the according graph gives the average iteration number of all 100,000 transformation cycles performed. In addition, for selected values of θ_k the standard deviation is marked by error bars. Results for arbitrary initial polygons are depicted in Figure 5.14a, those for valid initial polygons in Figure 5.14b. As can be seen, smaller angles result in a lower average number of iterations, since the eigenvalue ratios belonging to the other decomposition polygons are smaller. This can be seen by comparing the graph of iteration numbers with the graph of the eigenvalue ratios depicted in Figure 5.11. The standard deviation of the iteration numbers increases if θ increases. However, they differ only moderately for arbitrary and valid initial polygons.

Due to the use of random initial nodes, the probability of an initial polygon being invalid increases with its complexity, i.e., with its number of nodes. This is reflected by the valid initial polygon percentage amounting to 49.94% for triangles, 11.64% for quadrilaterals, and 5.62% for pentagons. Here, triangles are invalid if they are degenerated or clockwise-oriented, which, on average, is the case for each second triangle. For a larger number of nodes, the majority of all randomly generated elements are invalid and tangled. However, the transformation also reliably regularizes invalid elements, as can be seen by the average iteration numbers given in Figure 5.14. Furthermore, as is shown by the standard deviation error markers, the dispersion of iteration numbers is moderate.

Although in this chapter the regularizing polygon transformation has been defined and analyzed for planar polygons, it is also suitable for regularizing polygons in three-dimensional space, as will be considered in the following. The three-dimensional representation is derived similarly to the planar construction scheme depicted in Figure 5.4. Starting from a polygon P with $n \geq 3$ nodes $p_k \in \mathbb{R}^3$, $k \in \{0, \ldots, n-1\}$, for each edge $e(p_k, p_{k+1})$ a new node p'_{k+1} is constructed such that the edges $e(p_k, p_{k+1})$ and $e(p_k, p'_{k+1})$ enclose a predefined angle $\theta \in (0, \pi/2)$. Furthermore, the projection of p'_{k+1} subdivides $e(p_k, p_{k+1})$ in the point $m_k = \lambda p_k + (1-\lambda)p_{k+1}$. Thus, p'_{k+1} can be derived by rotating a scaled version of the edge $e(p_k, p_{k+1})$ around p_k by angle θ. Since the triangle $p_k p'_{k+1} m_k$ is right-angled, it holds that $\|m_k - p_k\|/\|p'_{k+1} - p_k\| = \cos\theta$. Using

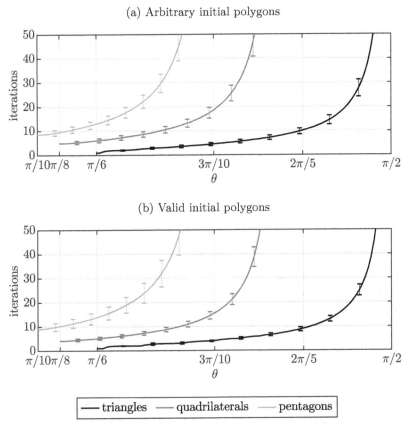

Figure 5.14: Average iteration numbers with respect to transformation parameters $\theta \in [\pi/(2n), 3\pi/(2n)]$, and $\lambda = 1/2$, required to regularize random polygons by the generalized polygon transformation.

that

$$\|m_k - p_k\| = \|\lambda p_k + (1 - \lambda)p_{k+1} - p_k\| = (1 - \lambda)\|p_{k+1} - p_k\|,$$

it follows that the length of the vector to be rotated by θ is given by

$$\|p'_{k+1} - p_k\| = \frac{\|m_k - p_k\|}{\cos\theta} = \frac{1 - \lambda}{\cos\theta}\|p_{k+1} - p_k\|.$$

This yields the basis for the following definition.

Definition 5.7 (Transformation of 3D polygons). *Let $w \in \mathbb{R}^3$ with $\|w\| = 1$ be a prescribed reference normal and P be a polygon with $n \geq 3$ nodes $p_k \in \mathbb{R}^3$, $k \in \{0, \ldots, n - 1\}$. The new nodes p'_k and p''_k obtained by the two consecutive substeps of the three-dimensional regularizing polygon transformation are* *defined by*

$$p'_k := p_{k-1} + \frac{1-\lambda}{\cos\theta} R(-w,\theta)(p_k - p_{k-1}) \quad and \tag{5.28}$$

$$p''_k := p'_{k+1} + \frac{1-\lambda}{\cos\theta} R(w,\theta)(p'_k - p'_{k+1}), \tag{5.29}$$

with the rotation matrix

$$R(w,\theta) := I + (\sin\theta)K(w) + (1 - \cos\theta)(K(w))^2,$$

based on the likewise 3×3 *matrices*

$$I := \begin{pmatrix} 1 & 0 & 0 \\ 0 & 1 & 0 \\ 0 & 0 & 1 \end{pmatrix} \quad and \quad K(w) := \begin{pmatrix} 0 & -w_2 & w_1 \\ w_2 & 0 & -w_0 \\ -w_1 & w_0 & 0 \end{pmatrix}$$

using zero-based indices for the coordinates w_k *of* w *for consistency reasons.*

The representations (5.28) and (5.29) follow directly by the construction scheme using Rodrigues' rotation formula for the rotation matrices $R(-w,\theta)$ and $R(w,\theta)$, representing a clockwise and counterclockwise rotation by θ, respectively. In both equations, the first node represents the rotation center and the rotation matrix is multiplied with the scaled oriented edge. The prescribed reference normal w is required to define the rotation, since the polygon nodes p_k do not have to be coplanar. In practice, such a reference normal is derived from additional geometric data, such as a local normal in a surface mesh. Using $w = (0,0,1)^t$ results in the same transformation scheme as the planar regularizing polygon transformation given by Lemma 5.12.

Figure 5.15 depicts the first seven polygons obtained by iteratively applying the regularizing polygon transformation according to Definition 5.7 for a three-dimensional hexagon with random nodes. Parameters of the transformation have been set to $w = (-1,-1,1)^t$, $\lambda = 1/2$, and $\theta = \pi/6$. The initial polygon depicted in the center is marked light gray. Subsequent iterate polygons are marked from darker shades of gray to black. Since no scaling has been used, iteratively applying the transformation increases the size of the resulting polygons. Figure 5.15 depicts the same result using two different perspectives: a front view shown on the left and a side view shown on the right.

As can be seen, due to its random nodes, the initial polygon is twisted in all three dimensions. Nevertheless, iteratively applying the polygon transformation seven times results in an almost regular hexagon with almost coplanar nodes. This is particularly visible in the side view depicted on the right. As can be seen, the transformed nodes iteratively move towards the plane defined by the centroid of the polygon and the prescribed normal w. In the example given, the maximal distance of the initial polygon nodes from this plane is 0.5236, which is reduced within seven transformation steps to 0.0852. As in the case of the planar transformation, the center of the polygon is preserved by the regularizing transformation.

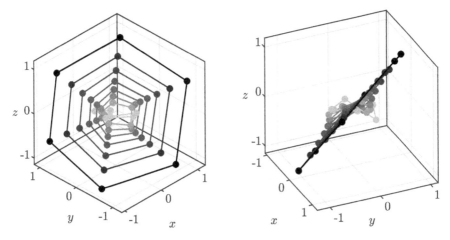

Figure 5.15: Iterative regularizing transformation of a random three-dimensional hexagon with reference normal $w = (-1, -1, 1)^t$, $\lambda = 1/2$, $\theta = \pi/6$ in front view (left) and side view (right).

The concept of regularizing polygons by iteratively applied transformations can also be generalized to non-Euclidean geometries, as is described in [144]. Following the spirit of using similar triangles, regularizing transformations are given for spheric and hyperbolic polygons. As in the case of the Euclidean plane described before, circulant transformation matrices play a key role in proving the regularizing effect of these transformations.

5.1.3 Customized polygon transformations

Depending on the application, regular elements might not have the ideal shape for mesh generation and smoothing. Examples for such applications are anisotropic meshes and thin boundary layers for computational fluid dynamics to support specific turbulence models.

Here, the analysis of the generalized polygon transformation based on the Fourier polygons given by Definition 5.3 and the associated eigenvalues provides a scheme for how to define customized polygon transformations with non-regular limits based on linear mappings. Applying such transformations iteratively to arbitrary initial polygons will result in polygons with prescribed *target shapes*. This will be demonstrated in the example of rectangle transformations leading to limit polygons with a prescribed side ratio.

As in the case of the regularizing transformation, the quadrilateral will be represented by a four-dimensional complex vector $z \in \mathbb{C}^4$ and the polygon transformation by a 4×4 matrix $M \in \mathbb{C}^{4 \times 4}$. In the case of the generalized regularizing polygon transformation, M has been derived from a geometric construction scheme. By analyzing the resulting circulant matrix M its eigenvalues and eigenvectors have been determined, which are directly linked to the

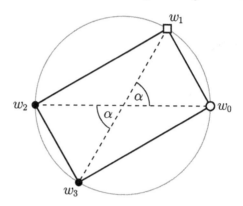

Figure 5.16: Customized target rectangle.

limit shape of the polygon obtained by iteratively applying the transformation. To be precise, the shape of the limit polygon depends on the dominating eigenvalue and the associated eigenvector. This knowledge is now readily being used to determine M for a prescribed target shape by an inverse process. That is, M will be defined by a suitable choice of eigenvectors and associated eigenvalues.

In the previous section, the eigenvectors turned out to be the Fourier polygons based on the roots of unity. Here, the nodes of the kth Fourier polygon are obtained by successively connecting counterclockwise each kth root of unity, starting from the first node one and scaling the resulting vector by $1/\sqrt{n}$. A similar approach will be used in the following. However, instead of taking roots of unity as eigenvector components, the nodes of a prescribed reference rectangle inscribed in the unit circle are used. This is illustrated in Figure 5.16.

The first node of the rectangle is defined to be one. The side ratio of the target rectangle can be set by choosing the associated angle $\alpha \in (0, \pi/2)$ enclosed by the two diagonals. In doing so, the nodes of the depicted rectangle result in

$$w_0 := 1, \quad w_1 := e^{i\alpha}, \quad w_2 := -1, \quad w_3 := e^{i(\pi+\alpha)}$$

where $e^{i\alpha} = \cos(\alpha) + i\sin(\alpha)$. In Figure 5.16 the node w_0 is marked by a circle, w_1 by a square, and the others by discs.

By iteratively taking each kth node for the k column starting from w_0 and normalizing by $1/\sqrt{4} = 1/2$, the matrix V of eigenvectors results in

$$V = \frac{1}{2} \begin{pmatrix} w_0 & w_0 & w_0 & w_0 \\ w_0 & w_1 & w_2 & w_3 \\ w_0 & w_2 & w_0 & w_2 \\ w_0 & w_3 & w_2 & w_1 \end{pmatrix}.$$

Here indices are zero-based and taken modulo four. The four column vectors of V are denoted as v_k, $k \in \{0, \ldots, 3\}$ from left to right. As in the case of the generalized regularizing polygon transformation, the first column results in the preservation of the centroid and the second column represents the target shape.

For the diagonal matrix $D := \mathrm{diag}(\eta_0, \ldots, \eta_3)$ of four prescribed eigenvalues η_k, the diagonalization of M analogous to (5.9) results in $M := VDV^{-1}$. The centroid is preserved, if the first eigenvalue is set to one. Since the second eigenvector represents the target polygon, the associated eigenvalue must be the dominant eigenvalue. To iteratively damp the influence of the remaining two eigenvectors, the associated eigenvalues are set to values smaller than one. This results in a choice of

$$\eta_0 := 1, \quad \eta_1 > 1, \quad \eta_2 < 1, \quad \eta_3 < 1.$$

Using only real-valued eigenvalues also eliminates the rotational effect of the transformation, which is preferred in a mesh smoothing context.

In Figure 5.16 the angle has been set to $\alpha = \pi/3$. Setting, for example, $\eta_1 = 3/2$ and $\eta_2 = \eta_3 = 1/2$ results in the transformation matrix

$$M = \frac{1}{8} \begin{pmatrix} 7 & 2 - \sqrt{3}\mathrm{i} & -1 & \sqrt{3}\mathrm{i} \\ 2 + \sqrt{3}\mathrm{i} & 7 & -\sqrt{3}\mathrm{i} & -1 \\ -1 & \sqrt{3}\mathrm{i} & 7 & 2 - \sqrt{3}\mathrm{i} \\ -\sqrt{3}\mathrm{i} & -1 & 2 + \sqrt{3}\mathrm{i} & 7 \end{pmatrix}. \tag{5.30}$$

In contrast to the regularizing transformation for quadrilaterals given by (5.6), this transformation is not banded and not circulant. Applying this transformation iteratively results in a limit rectangle with side length ratio $|w_1 - w_0|/|w_3 - w_0| = 1/\sqrt{3}$.

Due to the choice $\eta_1 = 3/2 > 1$, iteratively applying the transformation leads to a sequence of successively increasing polygons. Therefore, as in the case of the generalized regularizing polygon transformation, applying edge length preserving scaling after each transformation step is recommended.

Figure 5.17 depicts examples of iteratively applying the transformation (5.30) five times to a random initial quadrilateral $z^{(0)} \in \mathbb{C}^4$, leading to the sequence $z^{(\ell)} := M^\ell z^{(0)}$ with $\ell \in \{0, \ldots, 5\}$. The fast convergence is due to the specific choice of the eigenvalues, since iteratively applying the transformation leads to an exponential amplification of the part belonging to the target polygon and exponential damping of the parts belonging to the last two eigenvectors. A slower convergence can be obtained by selecting eigenvalues closer to one or by applying relaxation according to Definition 5.6.

Similar to the node-marking scheme of the target rectangle depicted in Figure 5.16, the first, second, and following nodes of each polygon are shown as circles, squares and discs, respectively, in Figure 5.17. Since the target rectangle is oriented counterclockwise, the resulting polygons are also oriented counterclockwise even if the initial rectangle is oriented clockwise, as can be seen in the first row. Furthermore, the centroid is preserved due to the specific

$z^{(0)}$ $z^{(1)}$ $z^{(2)}$ $z^{(3)}$ $z^{(4)}$ $z^{(5)}$

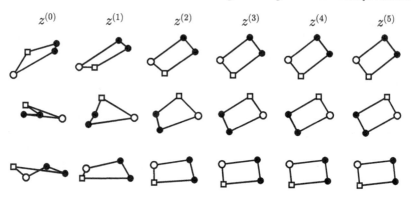

Figure 5.17: Examples of iteratively applying a customized transformation to random initial quadrilaterals.

choice of v_0 and η_0. However, the direction of the limit rectangle depends on the initial configuration, as is visible in the results depicted in the right column of Figure 5.17.

Depending on the application context, it might also be desirable to control the final direction, that is, the alignment of the element. This can be conducted by incorporating a controlled rotation into the transformation, which corrects the influence of the initial configuration of $z^{(0)}$. To do so, the polygon $z^{(\ell)}$ is represented first as a linear combination with respect to the basis vectors v_k using coefficients $u_k^{(\ell)}$, i.e.,

$$z^{(\ell)} = \sum_{k=0}^{3} u_k^{(\ell)} v_k = V u^{(\ell)}, \quad \text{with} \quad u^{(\ell)} := \left(u_0^{(\ell)}, \ldots, u_3^{(\ell)} \right)^{\mathrm{t}} \in \mathbb{C}^4.$$

It follows readily that $u^{(\ell)} = V^{-1} z^{(\ell)}$. Since v_1 represents the target shape, the absolute value $|u_1^{(\ell)}|$ and argument $\arg(u_1^{(\ell)})$ of the complex coefficient $u_1^{(\ell)}$ provide the scaling and rotation information with respect to this reference polygon.

In the following, the transformation is modified exemplarily in such a way that the limit rectangle is axis aligned and the first node is located on the lower right. That is, the limit shape is given by the rectangle depicted in Figure 5.16 rotated by angle $-\alpha/2$. Since $\arg(u_1^{(\ell)})$ represents the current angle of v_1 in $z^{(\ell)}$, the angle deviation with respect to the target angle $-\alpha/2$ amounts to $\arg(u_1^{(\ell)}) + \alpha/2$. To correct this deviation, an additional rotation by a fraction $\varrho \in (0, 1]$ of this angle deviation around the element centroid in the opposite direction is incorporated into the transformation. This rotation can be represented by multiplying with the complex number $\exp\left(-\varrho\mathrm{i}(\arg(u_1^{(\ell)}) + \alpha/2)\right)$, which results in the modified transformed nodes

$$z^{(\ell+1)} := e^{-\varrho\mathrm{i}(\arg(u_1^{(\ell)}) + \alpha/2)} \left(M z^{(\ell)} - c \right) + c, \tag{5.31}$$

$$z^{(0)} \qquad z^{(1)} \qquad z^{(2)} \qquad z^{(3)} \qquad z^{(4)} \qquad z^{(5)}$$

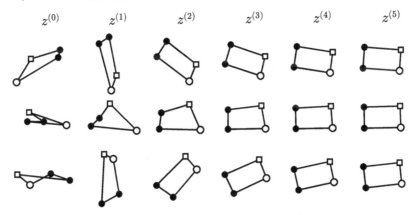

Figure 5.18: Examples of iteratively applying a customized transformation with direction correction to random initial quadrilaterals.

where c denotes the four-dimensional vector with each element set to the preserved initial element centroid $(1/4)\sum_{k=0}^{3} z_k^{(0)}$. Since this represents a complex scaling of the transformed element, this direction-correction step can naturally be combined with the edge length-preserving scaling step.

The result of applying the modified transformation (5.31) for the choice $\alpha = \pi/3$ and $\varrho = 1/2$ is depicted in Figure 5.18. Here, the same initial polygons have been used as shown in the left column of Figure 5.17. Again, the first and second nodes are marked by a circle and a square, respectively.

The aligning effect with respect to the given target angle $-\alpha/2$ of the transformation defined by (5.31) can be seen by comparing the results of the non-corrected transformation $z^{(\ell+1)} := Mz^{(\ell)}$ given in Figure 5.17 with those of the corrected transformation depicted in Figure 5.18. Here, the polygons shown in the right column are all oriented counterclockwise and have almost reached their prescribed limit given by an axis-aligned rectangle with the first node located on the lower right and an edge length ratio of $1/\sqrt{3}$.

Due to the choice $\varrho = 1/2$ in the given example, the angle deviation is halved by applying one step of the corrected transformation. Choosing larger values for ϱ results in a more effective correction of the angle deviation. If $\varrho = 1$ is used, the angle deviation is fully corrected by each transformation step.

It should also be noticed that the correction angle is adjusted for each step of the transformation to consider the current configuration. In doing so, the coefficient $u_1^{(\ell)} = (V^{-1}z^{(\ell)})_1$ must be computed. However, the iteration independent inverse V^{-1} has to be computed only once. In each iteration step, $u_1^{(\ell)}$ can then be derived by multiplying the second row of V^{-1} with $z^{(\ell)}$. For the given example, this results in

$$u_1^{(\ell)} := \left(\frac{1}{2}, \frac{1}{1+\sqrt{3}i}, -\frac{1}{2}, \frac{-1}{1+\sqrt{3}i} \right) z^{(\ell)}.$$

Following this construction scheme, customized transformations can easily be defined for polygons with an arbitrary number of nodes. By iteratively applying this transformation, arbitrary initial polygons can be transformed in prescribed limit shapes with a given orientation and direction.

5.2   Transformation of polyhedral elements

In the previous section, regularizing transformations have been presented for polygonal elements with an arbitrary number of nodes. All those elements have in common that each node has two incident edges. Since the number of nodes matches the number of edges, new nodes can be derived by constructing similar triangles on the edges of the polygon. This results in a unified construction scheme, which does not depend on the number of nodes, that is, on the polygonal element type.

In contrast, for polyhedral elements the number of incident edges per node depends on the element type. Furthermore, like in the case of the quadrilateral pyramid, an element can also consist of nodes with a different number of incident edges. Thus, geometric transformations for regularizing polyhedral elements will depend on the element type.

Nevertheless, as will be shown in the following, there is a unified construction principle for transforming the most common element types. This was first presented by the authors in [160]. However, before such transformations are introduced, a special transformation for tetrahedral elements first presented in [163] is considered, which has a particularly simple construction scheme.

5.2.1   Opposite face normals-based transformations

Let T denote a tetrahedron with the nodes $p_k \in \mathbb{R}^3$, $k \in \{1, \ldots, 4\}$, numbered according to Figure 2.2. It will be transformed by constructing on each node a scaled normal of the opposite face.

Definition 5.8 (Opposite face normal transformation). *For a non-degenerate tetrahedron T let the cross products*

$$n_1 := (p_4 - p_2) \times (p_3 - p_2), \qquad n_2 := (p_4 - p_3) \times (p_1 - p_3),$$
$$n_3 := (p_2 - p_4) \times (p_1 - p_4), \qquad n_4 := (p_2 - p_1) \times (p_3 - p_1)$$

denote the four face normals with Euclidean norms $\|n_k\|$. For a positive scaling factor $\sigma > 0$, the nodes p'_k of the transformed tetrahedron T' are given by

$$p'_k := p_k + \frac{\sigma}{\sqrt{\|n_k\|}} n_k, \tag{5.32}$$

where $k \in \{1, \ldots, 4\}$.

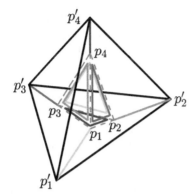

Figure 5.19: Opposite face normal transformation of a tetrahedron.

Applying this transformation, for example, to the tetrahedron T with nodes

$$p_1 = (0,0,0)^t, \quad p_2 = (1,0,0)^t, \quad p_3 = (0,2,0)^t, \quad p_4 = (0,0,3)^t$$

results in the tetrahedron T' with the nodes

$$p_1' = -\frac{\sigma}{\sqrt{7}}(6,3,2)^t, \qquad\qquad p_2' = (1 + \sqrt{6}\sigma, 0, 0)^t,$$

$$p_3' = (0, 2 + \sqrt{3}\sigma, 0)^t, \qquad\qquad p_4' = (0, 0, 3 + \sqrt{2}\sigma)^t.$$

This is depicted in Figure 5.19 for the specific choice $\sigma = 1$. Here, each face of T is marked by a gray, slightly shrunken triangle. The same shade of gray is used for the associated scaled normal constructed on the node opposite to each face. The initial tetrahedron with nodes p_k is marked with gray dashed lines, the transformed tetrahedron with nodes p_k' is marked black.

In the given example, the initial element mean ratio quality number $q(T) = 0.5943$ is improved to $q(T') = 0.9862$ by one step of the transformation. That is, T' is almost regular. However, there is no value of σ, which results in a regular tetrahedron after one transformation step, as will be shown in the following by a simple counterexample. This demonstrates that there is no analogon to Napoleon's theorem for triangular elements in the case of the opposite face normals transformation for tetrahedral elements.

To be regular, all edge lengths of the transformed tetrahedron must be equal. Since the equation $\|p_2' - p_3'\| = \|p_2' - p_4'\|$ has the only valid solution $\sigma = \sqrt{2} + \sqrt{3}$, and in contradiction $\|p_2' - p_3'\| = \|p_3' - p_4'\|$ implies $\sigma = (\sqrt{2} + \sqrt{6})/2$, there is no $\sigma > 0$ for which the tetrahedron T' obtained by one step of the transformation is regular. However, T can be regularized up to any given tolerance by applying several steps of the transformation.

Due to the orientation of the normals, the transformation enlarges valid tetrahedra and preserves their orientation. However, in contrast to the regularizing polygon transformations given in Section 5.1.2, the opposite face normals transformation does not preserve the centroid of the initial tetrahedron.

This can be seen for the given example with initial centroid $(0.25, 0.5, 0.75)^t$ but with the transformed tetrahedron centroid being approximately $(0.30, 0.65, 0.91)^t$ for the choice $\sigma = 1$. However, centroid preservation can easily be obtained by a subsequently applied translation step.

The translation of the centroid caused by the transformation is due to the scaling of the normals by the inverse of the square roots of their lengths. Not applying this normalization would have resulted in centroid preservation, since all unscaled normals sum up to the zero vector. However, scaling is essential to guarantee the following properties of the transformation, which are preferred in a mesh smoothing context.

Lemma 5.14 (Transformation invariance). *The transformation given by Definition 5.8 is invariant with respect to translation, rotation, and scaling.*

Proof. It has to be shown that for arbitrary scaling factors $\zeta > 0$, rotation matrices R, i.e., 3×3 matrices with $\det R = 1$ and $R^{-1} = R^t$, and translation vectors $t \in \mathbb{R}^3$, it holds that $(\zeta R p_k + t)' = \zeta R p'_k + t$.

Let $\tilde{p}_k := \zeta R p_k + t$ denote the affine transformation of p_k. Furthermore, let $F_{k,i}$ denote the index of the ith node of the kth face of the tetrahedron. The normals of these tetrahedron nodes under affine transformation are given by

$$
\begin{aligned}
\tilde{n}_k &:= \left(\tilde{p}_{F_{k,2}} - \tilde{p}_{F_{k,1}} \right) \times \left(\tilde{p}_{F_{k,3}} - \tilde{p}_{F_{k,1}} \right) \\
&= \left(\zeta R(p_{F_{k,2}} - p_{F_{k,1}}) \right) \times \left(\zeta R(p_{F_{k,3}} - p_{F_{k,1}}) \right) \\
&= \zeta^2 \det(R) \left(R^{-1} \right)^t \left[(p_{F_{k,2}} - p_{F_{k,1}}) \times (p_{F_{k,3}} - p_{F_{k,1}}) \right] = \zeta^2 R n_k,
\end{aligned}
$$

using that $\det(R) = 1$ and $(R^{-1})^t = R$.

Applying the opposite face normals transformation (5.32) to the nodes $\tilde{p}_k := \zeta R p_k + t$ of the tetrahedron under affine transformation implies

$$
\begin{aligned}
\tilde{p}'_k &= \tilde{p}_k + \frac{\sigma}{\sqrt{\|\tilde{n}_k\|}} \tilde{n}_k = \zeta R p_k + t + \frac{\sigma}{\zeta \sqrt{\|n_k\|}} \zeta^2 R n_k \\
&= \zeta R \left(p_k + \frac{\sigma}{\sqrt{\|n_k\|}} n_k \right) + t = \zeta R p'_k + t,
\end{aligned}
$$

which proves Lemma 5.14.   □

Due to the normalization used in the opposite face normals transformation and the resulting complexity of the involved expressions after transformation, there is yet no analytic proof of the regularizing property of the transformation given by (5.32). However, using a different scaling scheme results in a regularizing transformation for tetrahedral elements with a provable regularizing effect, as is described in Section 8.1.3.

Nevertheless, the regularizing property and the rate of regularization are demonstrated by the following results of an extensive numerical test like that of the polygonal case described in Section 5.1.2.3. The test is based on generating 100,000 random initial tetrahedra T_j and taking 101 equidistant values σ_k in

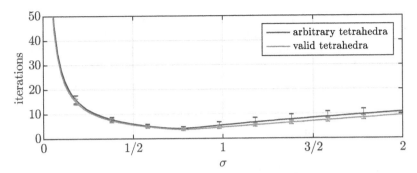

Figure 5.20: Average iteration numbers with respect to transformation parameter $\sigma \in [0,2]$ required to regularize random tetrahedra by the opposite face normals-based transformation.

$[0,2]$. For each pair (T_j, σ_k) taken from the Cartesian product of all elements and all σ-values, the opposite face normals transformation using the fixed scaling factor σ_k has been iteratively applied starting with the initial element T_j until the mean ratio quality number of the resulting element deviates less than 10^{-6} from the ideal value one. In doing so, for each pair (T_j, σ_k) a number of required iterations has been derived.

Generating a non-degenerate tetrahedron by randomly selected nodes, there is a 50% chance for the element to be valid, since the fourth node lies either on one or the other side of the plane defined by the other three nodes. In the given example, 50.15% of the arbitrarily chosen initial random elements are valid, while all other elements have a negative volume. As will be demonstrated by this test, element validity is not a necessary prerequisite for the opposite face normals transformation to result in regular tetrahedra. Nevertheless, the same test has also been conducted for 100,000 valid initial random tetrahedral elements for comparison reasons.

Figure 5.20 depicts the graphs for both tests, giving the average iteration number for a prescribed value of σ for all 100,000 initial elements to reach the prescribed element quality. In addition, for selected values of σ, the standard deviation is marked by error markers in both cases. As can be seen, element validity has only a minor impact on the iteration number required to result in an almost regular valid tetrahedron. Furthermore, due to the small standard deviation, the same holds for the initial element quality.

The figure also shows the influence of the scaling parameter σ on the rate of convergence. For very small values of σ the initial nodes and the associated transformed nodes differ only slightly, which results in a reduced regularizing effect. This is particularly useful in applications where a rapid change in the geometric configuration is undesirable. Similarly, iteration numbers are also increasing, if σ becomes large. According to the graphs, the lowest iteration numbers resulting in 3.69 and 4.09 are achieved for $\sigma \approx 0.78$ and $\sigma \approx 0.76$ in the case of all valid and arbitrary initial tetrahedra, respectively.

The approach of using opposite face normals is less obvious for other volumetric element types, such as pyramids or prisms. However, the next section describes a variation of the opposite face normals transformation being applicable to all volumetric element types under consideration. The key idea in transforming such elements is to use dual element face normals instead of using element face normals.

5.2.2 Dual element-based transformations

In the majority of cases, mixed volume meshes are assembled of tetrahedra, hexahedra, quadrilateral pyramids, or triangular prisms. Thus, this section focuses on such elements only. The geometric transformation schemes for these elements are based on using *dual elements* as given in the following.

Node numbering schemes for the element types under consideration as well as their dual elements are depicted in Figure 5.21. The dual elements, marked gray, consist of the nodes d_k, which will be defined based on the element faces. The node indices of each face are given by the entries of face node tuples defined by Equations (2.1) to (2.4).

The node d_k of a dual element is defined as the centroid of the kth element face, where face nodes do not necessarily have to be coplanar. I.e.,

$$d_k := \frac{1}{|F_k|} \sum_{i=1}^{|F_k|} p_{F_{k,i}}, \quad k \in \{1, \dots, |F|\}, \tag{5.33}$$

where $|F_k|$ denotes the number of nodes in the kth face and $|F|$ the number of element faces. For example, in the case of the pyramid it holds that $|F^{\text{pyr}}| = 5$, $|F_1| = 4$, and $|F_2| = |F_3| = |F_4| = |F_5| = 3$, since this element consists of one quadrilateral face and four triangular faces.

The tetrahedron and the pyramid are self-dual, that is, their dual elements are likewise a tetrahedron and a pyramid, respectively. In contrast, the dual elements of the hexahedron and the prism are an octahedron and a triangular dipyramid, respectively. In addition, among all dual element faces, the base quadrilateral of the dual pyramid is the only non-triangular face. The following defines the *dual element faces tuples*, i.e., node indices of the faces of all dual elements:

$\bar{F}^{\text{tet}} := \big((1,2,4),(1,3,2),(1,4,3),(2,3,4)\big),$

$\bar{F}^{\text{hex}} := \big((1,2,5),(1,3,2),(1,4,3),(1,5,4),(6,5,2),(6,2,3),(6,3,4),(6,4,5)\big),$

$\bar{F}^{\text{pyr}} := \big((1,2,5),(1,3,2),(1,4,3),(1,5,4),(2,3,4,5)\big),$

$\bar{F}^{\text{pri}} := \big((1,2,4),(1,3,2),(1,4,3),(5,4,2),(5,2,3),(5,3,4)\big).$

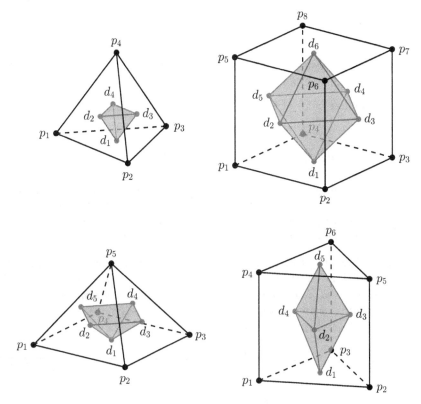

Figure 5.21: Node numbering schemes for all element types marked black and their duals marked gray.

The regularizing transformation scheme for polyhedral elements is based on constructing scaled versions of the dual element face normals on suitable base points. Here, the outward directed face normals are defined by the cross products

$$
n_k := \begin{cases} (d_{\bar{F}_{k,2}} - d_{\bar{F}_{k,1}}) \times (d_{\bar{F}_{k,3}} - d_{\bar{F}_{k,1}}) & \text{if } |\bar{F}_k| = 3 \\ \frac{1}{2}(d_{\bar{F}_{k,3}} - d_{\bar{F}_{k,1}}) \times (d_{\bar{F}_{k,4}} - d_{\bar{F}_{k,2}}) & \text{if } |\bar{F}_k| = 4, \end{cases} \tag{5.34}
$$

where $k \in \{1, \dots, |\bar{F}|\}$ and $\bar{F}$ denotes a dual element face tuple taken from $\{\bar{F}^{\text{tet}}, \bar{F}^{\text{hex}}, \bar{F}^{\text{pyr}}, \bar{F}^{\text{pri}}\}$. That is, in the case of triangular faces, the normal is determined by the cross product of two vectors spanning the face. In the case of the quadrilateral base of the pyramid, the normal is defined as the cross product of the two diagonals of the quadrilateral scaled by $1/2$. This normal is equal to the arithmetic mean of the normals of the four triangles, each defined by three consecutive points of the quadrilateral face.

The base points, on which the normals will be erected, are defined with the aid of the dual element face centroids given by

$$c_k := \frac{1}{|\bar{F}_k|} \sum_{i=1}^{|\bar{F}_k|} d_{\bar{F}_{k,i}}, \quad k \in \{1, \ldots, |\bar{F}|\}. \tag{5.35}$$

For the triangular faces of the duals of non-Platonic elements, i.e., pyramids and prisms, the point

$$a_k(\tau) := (1 - \tau)d_{\bar{F}_{k,1}} + \tau \frac{1}{2}\left(d_{\bar{F}_{k,2}} + d_{\bar{F}_{k,3}}\right) \tag{5.36}$$

with $\tau \in \mathbb{R}$ is used. It is located on the line connecting the triangular face apex and the midpoint of the opposing side. The following definition specifies the choice of τ with respect to a normals scaling factor $\sigma > 0$. The latter will be used to control the regularizing effect of the single element transformation.

Definition 5.9 (Dual element-based transformations for polyhedra). *Let $E \in \{E^{\text{tet}}, E^{\text{hex}}, E^{\text{pyr}}, E^{\text{pri}}\}$ denote an arbitrary polyhedral element with $|E|$ nodes $p_k \in \mathbb{R}^3$ and dual element face normals n_k. Furthermore, let $\sigma > 0$ denote an arbitrary normals scaling factor. The nodes p'_k of the transformed element $E' = (p'_1, \ldots, p'_{|E|})$ are given by*

$$p'_k := b_k + \frac{\sigma}{\sqrt{\|n_k\|}} n_k, \quad k \in \{1, \ldots, |E|\}, \tag{5.37}$$

with element type-dependent base points

$$b_k := \begin{cases} c_k & \text{if } E \in \{E^{\text{tet}}, E^{\text{hex}}\} \text{ or} \\ & \quad (E = E^{\text{pyr}} \text{ and } k = 5) \\ a_k(\tau), \ \tau := \dfrac{1}{2} + \sigma & \text{if } E = E^{\text{pyr}} \text{ and } 1 \le k \le 4 \\ a_k(\tau), \ \tau := \dfrac{4}{5}\left(1 - \dfrac{\sqrt{2}\sigma}{\sqrt[4]{39}}\right) & \text{if } E = E^{\text{pri}}. \end{cases} \tag{5.38}$$

The construction of the transformed element is depicted in Figure 5.22 for the specific choice $\sigma = 1$. Here, the faces of the dual elements are marked gray. The base points b_k are indicated by black markers, the normals n_k by gray arrows, and the edges of the resulting transformed elements E' are marked black. In the case of the pyramidal and prismatic element, the lines connecting each triangular face apex with the midpoint of the opposite side are marked by dashed lines. Here, the resulting base point b_k can be located outside the associated dual element face, as can be seen in the case of the pyramidal element.

A basic prerequisite for regularizing element transformations is that regular elements are transformed into regular elements. For tetrahedral and hexahedral elements this holds, due to symmetry reasons, for the specific choice of $b_k = c_k$ and arbitrary values of $\sigma > 0$. In contrast, using c_k as base point does not

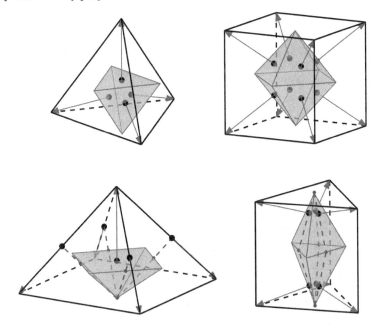

Figure 5.22: Transformed element construction by erecting normals on the dual element.

preserve element regularity in the case of the non-Platonic pyramidal and prismatic elements. Hence, for these elements an additional degree of freedom had to be included. It is represented by the σ-dependent parameter τ defining the position of the base points b_k of triangular faces chosen on the line connecting the apex and the midpoint of the opposing edge. The specific choice of τ given in (5.38) has been determined by the basic regularity prerequisite. That is, by solving an equation based on the equality of transformed element edge lengths.

This computation will now be given exemplarily for the regular quadrilateral pyramid with the nodes

$$p_1 = \begin{pmatrix} 0 \\ 0 \\ 0 \end{pmatrix}, \quad p_2 = \begin{pmatrix} 1 \\ 0 \\ 0 \end{pmatrix}, \quad p_3 = \begin{pmatrix} 1 \\ 1 \\ 0 \end{pmatrix},$$

$$p_4 = \begin{pmatrix} 0 \\ 1 \\ 0 \end{pmatrix}, \quad p_5 = \begin{pmatrix} 1/2 \\ 1/2 \\ 1/\sqrt{2} \end{pmatrix} \tag{5.39}$$

and all edges having the length one. According to (5.33) the pyramid face centroids representing the nodes of the dual pyramid are given by

$$d_1 = \begin{pmatrix} 1/2 \\ 1/2 \\ 0 \end{pmatrix}, \quad d_2 = \begin{pmatrix} 1/2 \\ 1/6 \\ 1/(3\sqrt{2}) \end{pmatrix}, \quad d_3 = \begin{pmatrix} 5/6 \\ 1/2 \\ 1/(3\sqrt{2}) \end{pmatrix},$$

$$d_4 = \begin{pmatrix} 1/2 \\ 5/6 \\ 1/(3\sqrt{2}) \end{pmatrix}, \quad d_5 = \begin{pmatrix} 1/6 \\ 1/2 \\ 1/(3\sqrt{2}) \end{pmatrix}.$$

Following (5.36), the base nodes a_k depending on the parameter τ are given by

$$a_1 = \begin{pmatrix} (3-\tau)/6 \\ (3-\tau)/6 \\ \tau/(3\sqrt{2}) \end{pmatrix}, \quad a_2 = \begin{pmatrix} (3+\tau)/6 \\ (3-\tau)/6 \\ \tau/(3\sqrt{2}) \end{pmatrix}, \quad a_3 = \begin{pmatrix} (3+\tau)/6 \\ (3+\tau)/6 \\ \tau/(3\sqrt{2}) \end{pmatrix},$$

$$a_4 = \begin{pmatrix} (3-\tau)/6 \\ (3+\tau)/6 \\ \tau/(3\sqrt{2}) \end{pmatrix}, \quad a_5 = \begin{pmatrix} 1/2 \\ 1/2 \\ 1/(3\sqrt{2}) \end{pmatrix}.$$

The nodes of the transformed pyramid are obtained by adding the scaled face normals $\hat{n}_k := (1/\sqrt{\|n_k\|})n_k$, given by

$$\hat{n}_1 = \begin{pmatrix} -1/(3\sqrt[4]{8}) \\ -1/(3\sqrt[4]{8}) \\ -1/(3\sqrt[4]{2}) \end{pmatrix}, \quad \hat{n}_2 = \begin{pmatrix} 1/(3\sqrt[4]{8}) \\ -1/(3\sqrt[4]{8}) \\ -1/(3\sqrt[4]{2}) \end{pmatrix}, \quad \hat{n}_3 = \begin{pmatrix} 1/(3\sqrt[4]{8}) \\ 1/(3\sqrt[4]{8}) \\ -1/(3\sqrt[4]{2}) \end{pmatrix},$$

$$\hat{n}_4 = \begin{pmatrix} -1/(3\sqrt[4]{8}) \\ 1/(3\sqrt[4]{8}) \\ -1/(3\sqrt[4]{2}) \end{pmatrix}, \quad \hat{n}_5 = \begin{pmatrix} 0 \\ 0 \\ \sqrt{2}/3 \end{pmatrix},$$

with n_k defined by (5.34) and multiplied by $\sigma > 0$ to the base nodes a_k, i.e., $p'_k = a_k + \sigma \hat{n}_k$. This implies, for example,

$$p'_1 = \begin{pmatrix} (3-\tau)/6 \\ (3-\tau)/6 \\ \tau/(3\sqrt{2}) \end{pmatrix} + \sigma \begin{pmatrix} -1/(3\sqrt[4]{8}) \\ -1/(3\sqrt[4]{8}) \\ -1/(3\sqrt[4]{2}) \end{pmatrix},$$

$$p'_2 = \begin{pmatrix} (3+\tau)/6 \\ (3-\tau)/6 \\ \tau/(3\sqrt{2}) \end{pmatrix} + \sigma \begin{pmatrix} 1/(3\sqrt[4]{8}) \\ -1/(3\sqrt[4]{8}) \\ -1/(3\sqrt[4]{2}) \end{pmatrix},$$

$$p'_5 = \begin{pmatrix} 1/2 \\ 1/2 \\ 1/(3\sqrt{2}) \end{pmatrix} + \sigma \begin{pmatrix} 0 \\ 0 \\ \sqrt{2}/3 \end{pmatrix}.$$

As mentioned before, as a prerequisite, a regularizing transformation must transform a regular pyramid into a regular pyramid. This implies that all edge lengths must be equal. Due to symmetry reasons it suffices to choose τ such that $\|p_2' - p_1'\| = \|p_5' - p_1'\|$ is fulfilled independent of the choice of the scaling factor $\sigma > 0$. Using the representation of the transformed nodes given before, this results in the equation

$$\frac{1}{3}\left|\tau + \sqrt[4]{2}\sigma\right| = \frac{1}{3\sqrt{2}}\sqrt{\left(\tau + \sqrt[4]{2}\sigma\right)^2 + \left(1 - \tau + (2 + \sqrt[4]{2})\sigma\right)^2}.$$

Solving for τ yields the solution $\tau = \sigma + 1/2$ as used by Definition 5.9. The choice of τ for the prism follows by an analog computation.

Lemma 5.15 (Transformation invariance). *The transformation given by Definition 5.9 is invariant with respect to translation, rotation, and scaling for all element types.*

Proof. It has to be shown that for arbitrary scaling factors $\zeta > 0$, rotation matrices R, i.e., 3×3 matrices with $\det R = 1$ and $R^{-1} = R^t$, and translation vectors $t \in \mathbb{R}^3$, it holds that $(\zeta R p_k + t)' = \zeta R p_k' + t$. This will be accomplished by applying the geometric transformation given by Definition 5.9 to the nodes $\tilde{p}_k := \zeta R p_k + t$ of an element $E \in \{E^{\text{tet}}, E^{\text{hex}}, E^{\text{pyr}}, E^{\text{pri}}\}$ with the corresponding faces tuple F.

According to (5.33) the nodes $\tilde{d}_k$ of the dual element associated to the element with nodes $\tilde{p}_k$ are given by

$$\tilde{d}_k := \frac{1}{|F_k|}\sum_{i=1}^{|F_k|}\tilde{p}_{F_{k,i}} = \frac{1}{|F_k|}\sum_{i=1}^{|F_k|}(\zeta R p_{F_{k,i}} + t)$$

$$= \zeta R\left(\frac{1}{|F_k|}\sum_{i=1}^{|F_k|}p_{F_{k,i}}\right) + t = \zeta R d_k + t,$$

with $k \in \{1, \ldots, |F|\}$. Using (5.35) it can be shown analogously that the centroid $\tilde{c}_k$ of the kth dual element face with node indices given by $\bar{F}_k$ can be written as

$$\tilde{c}_k := \frac{1}{|\bar{F}_k|}\sum_{i=1}^{|\bar{F}_k|}\tilde{d}_{\bar{F}_{k,i}} = \zeta R c_k + t, \quad \text{for } k \in \{1, \ldots, |\bar{F}|\}.$$

According to (5.38), the base node $\tilde{b}_k$ is either the centroid $\tilde{c}_k$ of the kth dual element face or given by the linear combination

$$\tilde{a}_k(\tau) := (1 - \tau)\tilde{d}_{\bar{F}_{k,1}} + \frac{\tau}{2}\left(\tilde{d}_{\bar{F}_{k,2}} + \tilde{d}_{\bar{F}_{k,3}}\right)$$

$$= (1 - \tau)(\zeta R d_{\bar{F}_{k,1}} + t) + \frac{\tau}{2}\left(\zeta R(d_{\bar{F}_{k,2}} + d_{\bar{F}_{k,3}}) + 2t\right)$$

$$= \zeta R\left((1 - \tau)d_{\bar{F}_{k,1}} + \frac{\tau}{2}\left(d_{\bar{F}_{k,2}} + d_{\bar{F}_{k,3}}\right)\right) + t = \zeta R a_k(\tau) + t,$$

which implies $\tilde{b}_k = \zeta R b_k + t$ for $k \in \{1, \ldots, |E|\}$. That is, the normal base nodes of the scaled, rotated, and translated element can be obtained by scaling, rotating, and translating the normal base nodes of the initial element. The remaining equalities $\tilde{n}_k = \zeta^2 R n_k$ and finally $\tilde{p}'_k = (\zeta R p_k + t)' = \zeta R p'_k + t$ follow analogously to the proof of Lemma 5.14. $\qquad\qquad\square$

The proof of Lemma 5.15 also gives the motivation for the choice of the normalization factor $1/\sqrt{\|n_k\|}$ in the definition (5.37) of p'_k, since it ensures the scale invariance of the regularizing element transformation.

For an arbitrary element $E \in \{E^{\text{tet}}, E^{\text{hex}}, E^{\text{pyr}}, E^{\text{pri}}\}$ with $|E|$ nodes p_k, let

$$q := \frac{1}{|E|} \sum_{k=1}^{|E|} p_k \tag{5.40}$$

denote the centroid of the initial element. Furthermore, let

$$q' := \frac{1}{|E|} \sum_{k=1}^{|E|} p'_k = \underbrace{\frac{1}{|E|} \sum_{k=1}^{|E|} b_k}_{=: q'_b} + \underbrace{\frac{\sigma}{|E|} \sum_{k=1}^{|E|} \frac{1}{\sqrt{\|n_k\|}} n_k}_{=: q'_n} \tag{5.41}$$

denote the centroid of the transformed element and its partition into the mean base node q'_b as well as the mean normal q'_n. The following lemma considers the relation between the centroids of initial and transformed elements.

Lemma 5.16 (Centroid of transformed elements). *For $E \in \{E^{\text{tet}}, E^{\text{hex}}, E^{\text{pri}}\}$ it holds that $q = q'_b$. In addition, for $E = E^{\text{hex}}$ it holds that $q = q'$, i.e., the transformation preserves the centroid of hexahedral elements.*

Proof. The hexahedral case will be shown first by evaluating the two sums q'_b and q'_n according to (5.41) separately. For q'_b, successively substituting $b_k := c_k$ by the representations of the octahedron face centroids and nodes yields

$$\frac{1}{8} \sum_{k=1}^{8} c_k = \frac{1}{8} \sum_{k=1}^{8} \left(\frac{1}{3} \sum_{i=1}^{3} \left(\frac{1}{4} \sum_{j=1}^{4} p_{F^{\text{hex}}_{\bar{F}^{\text{hex}}_{k,i}, j}} \right) \right)$$

$$= \frac{1}{8 \cdot 3 \cdot 4} \sum_{k=1}^{8} 3 \cdot 4 \cdot p_k = \frac{1}{8} \sum_{k=1}^{8} p_k.$$

The simplification of the triple sum is based on the fact that each node p_k is involved in three face centroids of the hexahedron and with each of these in four face centroids of the octahedron.

To prove $q = q'$, it remains to show that $q'_n = (0, 0, 0)^{\text{t}}$, which holds since the normals of opposing octahedron faces pairwise sum up to the zero vector. That is, $n_1 = -n_7$, $n_2 = -n_8$, $n_3 = -n_5$, and $n_4 = -n_6$, as will be shown exemplarily for the first identity. The representation of the octahedron normals implies

$$n_1 = (d_2 - d_1) \times (d_5 - d_1)$$

$$= +\frac{1}{16}\left[(p_5 + p_6 - p_3 - p_4) \times (p_5 + p_8 - p_2 - p_3)\right]$$

$$= -\frac{1}{16}\left[(p_2 + p_3 - p_5 - p_8) \times (p_3 + p_4 - p_5 - p_6)\right]$$

$$= -\left((d_3 - d_6) \times (d_4 - d_6)\right) = -n_7,$$

since the cross product is compatible with scalar multiplication and anti-commutative. The other identities can be deduced analogously.

In the case of tetrahedral elements, successively substituting the expressions (5.35) and (5.33) in the representation of the base nodes $b_k = c_k$ implies

$$q_b' = \frac{1}{4}\sum_{k=1}^{4} c_k = \frac{1}{4}\sum_{k=1}^{4}\left(\frac{1}{3}\sum_{i=1}^{3}\left(\frac{1}{3}\sum_{j=1}^{3} p_{\bar{F}_{k,i}^{tet},j}\right)\right) = \frac{1}{36}\sum_{k=1}^{4} 9p_k = q.$$

Here, each node p_k is involved in the three adjacent element face centroids d_k, and with each of these in three dual element face centroids c_k.

Finally, in the case of prismatic elements, the base nodes are given by $b_k = a_k(\tau)$, resulting in

$$q_b' = \frac{1}{6}\sum_{k=1}^{6} a_k(\tau) = \frac{1}{6}\sum_{k=1}^{6}\left[(1-\tau)d_{\bar{F}_{k,1}} + \frac{\tau}{2}\left(d_{\bar{F}_{k,2}} + d_{\bar{F}_{k,3}}\right)\right]$$

$$= \frac{1}{6}\left[3(1-\tau)(d_1 + d_5) + 2\tau(d_2 + d_3 + d_4)\right]$$

$$= \frac{1}{6}\left[(1-\tau)(p_1 + \cdots + p_6) + \tau(p_1 + \cdots + p_6)\right] = q.$$

Since $q_b' = q$ has been shown for arbitrary τ, this holds in particular for the specific choice of τ given in (5.38). $\qquad\square$

Whereas Lemma 5.16 implies $q_n' = (0,0,0)^t$ for arbitrary hexahedra, this does not generally hold for tetrahedral, pyramidal, and prismatic elements, as can be shown by simple counterexamples. For example, for the regular quadrilateral pyramid with nodes given by (5.39) and $\sigma = 1$, it holds that $q_n' = (0,0,(\sqrt{2} - 2\sqrt[4]{8})/15)^t$. However, centroid preservation can easily be achieved by an additional translation step after transforming the element. According to Lemma 5.16, the required translation vector is given by $-q_n'$ in the case of tetrahedral and prismatic elements and by $q - q'$ in the case of pyramidal elements. For the latter, even $q = q_b'$ does not hold in general. For example, the centroid of the regular quadrilateral pyramid with nodes (5.39) is given by $q = (1/2, 1/2, \sqrt{2}/10)^t$, which differs from the arithmetic mean $q_b' = (1/2, 1/2, 7\sqrt{2}/30)^t$ of the base nodes in the case $\sigma = 1$.

Due to its geometric construction and depending on the choice of the normals scaling factor σ given in Definition 5.9, the size of a transformed element can differ significantly from the size of the initial element. Analogous to the case of polygonal elements, this can be avoided by an element scaling step, which can be combined with the centroid preservation step.

Definition 5.10 (Transformed polyhedron scaling and centroid preservation). *Let $Q = (q, \ldots, q)$ and $Q' = (q', \ldots, q')$ denote $3 \times |E|$ matrices, with each column representing the initial and transformed element centroid q and q' according to (5.40) and (5.41), respectively. For an arbitrary scaling factor $\zeta > 0$, the scaled transformed element E'_s preserving the initial element centroid q is given by*

$$E'_s := Q + \zeta(E' - Q'). \tag{5.42}$$

E'_s is called an edge length scaled transformed polyhedron, *if $\zeta := \mathrm{el}(E) / \mathrm{el}(E')$, with $\mathrm{el}(E)$ denoting the sum of all edge lengths of the polyhedron E.*

For each of the four element types, exemplaric valid initial elements defined by randomly chosen nodes and their mean ratio quality numbers can be seen in the left column of Figure 5.23. Here, each row contains the elements obtained by iteratively applying the edge length scaled regularizing polyhedron transformation according to (5.42) to prevent the successive growth of the elements. In all cases, the transformation parameter $\sigma = 1$ has been used.

Mean ratio quality numbers are successively improved resulting in nearly regular elements after applying the transformation five times, as can be seen by the elements depicted in the right column. In particular, the first transformation step leads to a considerable improvement towards element regularity and with this in element quality. In the context of mesh smoothing, such a rapid change of shape is not always favorable since it also increases the risk of invalid neighbor element generation. Therefore, the same approach as in the case of the polygon relaxation scheme given by Definition 5.6 is introduced for polyhedra.

Definition 5.11 (Polyhedron relaxation). *Let E denote a polyhedral element and E'_s denote its transformed and scaled counterpart according to (5.42). The nodes of the* relaxed edge length scaled transformed polyhedron E'_r *are given by the columns of the linear combination*

$$E'_r := (1 - \varrho)E + \varrho E'_s$$

using a relaxation factor $\varrho \in [0, 1]$.

The smaller ϱ gets, the smaller are the geometric changes compared to the initial element E. For the choice $\varrho = 1$ it holds that $E'_r = E'_s$.

As has been mentioned in the introduction of this section, the polyhedral elements under consideration differ in their number of nodes and incident edges. Although Definition 5.9 of the transformation for polyhedra is based on a dual element approach for each element type, it is still unknown if there

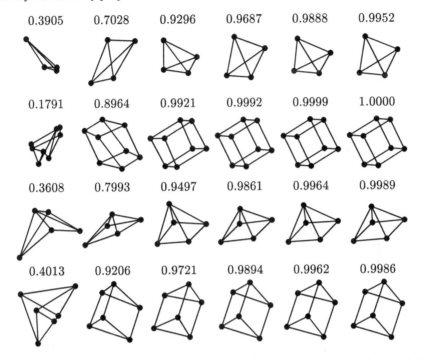

0.3905 0.7028 0.9296 0.9687 0.9888 0.9952

0.1791 0.8964 0.9921 0.9992 0.9999 1.0000

0.3608 0.7993 0.9497 0.9861 0.9964 0.9989

0.4013 0.9206 0.9721 0.9894 0.9962 0.9986

Figure 5.23: Elements and their mean ratio quality numbers obtained by successively applying the regularizing transformation using $\sigma = 1$ and mean edge length-preserving scaling.

is a suitable common representation for the analysis of the regularizing effect of this transformation comparable to the discrete Fourier matrix used in the analysis of the generalized polygon transformation. Furthermore, since cross products and vector normalizations are involved, the transformation given by (5.37) is non-linear. This is the reason why there is as yet no analytic proof of the regularizing effect of the polyhedron transformation. However, it has been substantiated by numerical tests, which also provide more insight into the regularizing behavior of the transformations with respect to the scaling parameter σ.

In accordance with the case of the polygon transformation and the opposite face normal transformation for tetrahedral elements, the tests are based on generating 100,000 random initial elements of each type and taking 101 equidistant values for $\sigma \in [0, 2]$. For each pair (E_j, σ_k) taken from the Cartesian product of all elements and all σ-values, the geometric transformation using the scaling factor σ_k has been iteratively applied starting with the initial element E_j until the mean ratio quality number of the resulting element deviates less than 10^{-6} from the ideal value one. After each iteration step the resulting element has been scaled with $\zeta := \mathrm{el}(E)/\mathrm{el}(E')$ to avoid numerical instabilities caused by the successive element growth, which would have occurred otherwise. Due to

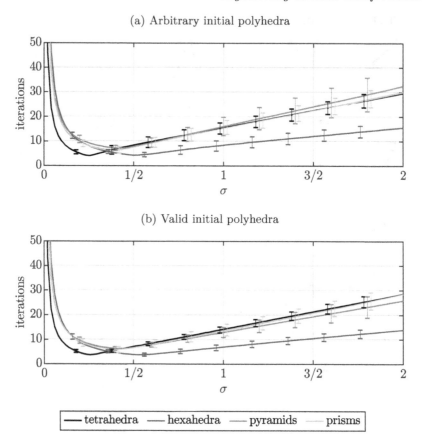

Figure 5.24: Average iteration numbers with respect to transformation parameter $\sigma \in [0, 2]$ required to regularize random polyhedra by the dual element-based transformations.

the scaling invariance of the transformation according to Lemma 5.15, this has no influence on the rate of regularization. In addition, relaxation has not been applied due to the specific choice $\varrho = 1$. For each element type, this test results in 10,100,000 transformation cycles, each represented by its number of iterations, thus providing an indication of the rate of regularization.

Results of this test are depicted in Figure 5.24a. Here, for each element type and specific choice of σ_k, the according graph gives the average iteration number of all 100,000 transformation cycles performed. In addition, for selected values of σ the standard deviation is marked by error markers. As can be seen, all graphs show a similar behavior. To be precise, the average iteration number increases if σ tends to zero, since the transformation tends to become a node-averaging scheme without a regularizing effect. For $\sigma \in (1/2, 2)$, that is after reaching its minimum, the average iteration number depends almost linearly on σ. Thus, the parameter σ provides an easy control for the rate of regularization.

Due to the usage of random initial nodes, the probability of an initial element being invalid increases with its complexity, i.e., with its number of nodes. This is reflected by the valid initial element percentage amounting to 50.15% for tetrahedral, 0.37% for hexahedral, 9.36% for pyramidal, and 5.33% for prismatic elements. That is, in the majority of cases, random generated elements are invalid. However, the transformation also reliably regularizes invalid elements, as can be seen by the average iteration numbers given in Figure 5.24a. Furthermore, as is shown by the standard deviation error markers, the dispersion of iteration numbers is moderate.

Regularization results are even more uniform if only valid initial elements are used, as is common in the case of mesh smoothing. Results based on using 100,000 different valid initial elements for each element type are depicted in Figure 5.24b. Compared to the test case of arbitrary initial elements, the average iteration number becomes smaller while preserving its qualitative behavior with respect to σ. Furthermore, the standard deviation is decreased. That is, for a given element type and normals scaling factor σ_k, elements are regularized quite uniformly. This is an important fact with respect to the mesh smoothing approach presented in the following chapter, assuring a consistent smoothing result.

The transformation of hexahedral elements not only differs in the centroid preservation property, but also in its increased rate of regularization, as can be seen in both cases. Hence, for mixed mesh smoothing, it is advisable to use element-type dependent transformation parameters to obtain a uniform regularization of the elements under consideration, as is the case for mixed polygonal mesh smoothing.

6

The GETMe smoothing framework

In the previous chapter, regularizing element transformations for polygonal and polyhedral elements have been considered. If applied iteratively, these transformations convert arbitrary single elements into their regular counterparts. Such transformations will now be used for defining a framework of smoothing method variants, providing a purely geometric approach to high-quality mesh smoothing. The resulting methods, called *geometric element transformation methods*, abbreviated as *GETMe*, are highly effective and efficient, although conceptually simple. GETMe variants are introduced in the following sections, starting with a basic scheme not involving any quality aspects, thus being entirely based on geometric operations. After that, refined versions are described, which incorporate a quality criterion for improved smoothing and termination control. Such methods are of particular interest in the context of finite element mesh smoothing that requires the resulting mesh to be valid.

6.1 Building blocks of the GETMe framework

6.1.1 Basic GETMe simultaneous smoothing

In this section the *basic GETMe simultaneous smoothing* approach will be introduced. Here, "basic" means that it is a purely geometry-driven smoothing approach not incorporating any quality criteria. The only ingredients required are regularizing transformations for each element type of the mesh. Like in the case of the basic Laplacian smoothing approach, new nodes are computed by a node averaging process. However, in contrast to Laplacian smoothing, this is not done by averaging neighboring nodes, but by averaging intermediate nodes obtained by the regularizing element transformations. The basic idea is depicted in Figure 6.1.

The mesh consists of $n_p = 7$ nodes $p_i \in \mathbb{R}^2$ and the $n_e = 4$ polygonal elements given by $E_1 = (p_1, p_2, p_3, p_4)$, $E_2 = (p_1, p_4, p_5)$, $E_3 = (p_1, p_5, p_6, p_7)$, and $E_4 = (p_1, p_7, p_2)$, as is depicted on the left of Figure 6.1.

First, each of the elements is transformed using the generalized polygon transformation, and scaled to preserve the edge length sum of its non-transformed counterpart. This results in the polygons shown in different shades

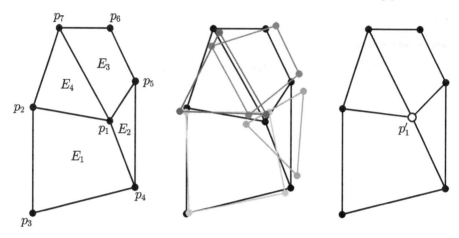

Figure 6.1: Node update scheme of basic GETMe simultaneous smoothing.

of gray, which are depicted in the middle of Figure 6.1. As can be seen, for each element, the transformation leads to a slightly modified position of the common node p_1.

After that, the new position p_1' of this node obtained by one iteration of the basic GETMe simultaneous smoothing scheme is defined as the arithmetic mean of the four transformed versions of p_1. The resulting mesh is depicted on the right. Here, the boundary nodes $p_2, \ldots, p_7$ have been kept fixed. Thus, p_1' is the only modified node. As can be seen, the node p_1' marked by a circle is moved towards the mesh center, thus improving mesh quality. The word simultaneous in the name of this smoothing scheme indicates that within one smoothing iteration all mesh elements are transformed simultaneously.

In the following, a more precise definition of one mesh smoothing iteration step is given. Let $I := \{1, \ldots, n_p\}$ denote the index set of all mesh nodes p_i and $J := \{1, \ldots, n_e\}$ denote the index set of all mesh elements E_j. Each mesh element E_j is defined by its nodes p_i with i taken from the element-specific node index set $I(j) \subseteq I$.

Transforming and edge length-preserving scaling of an element E_j leads for each of its nodes p_i to a new node position denoted as $p_{i,j}'$. Conversely, for each mesh node p_i and its incident elements E_j with j taken from the node-specific incident element index set $J(i) \subseteq J$, this results in $|J(i)|$ new temporary nodes $p_{i,j}'$. If p_i is a fixed mesh node, it is preserved. Otherwise p_i' is derived as the arithmetic mean of its associated transformed versions, i.e.,

$$p_i' := \begin{cases} p_i & \text{if } p_i \text{ is fixed} \\ \frac{1}{|J(i)|} \sum_{j \in J(i)} p_{i,j}' & \text{otherwise.} \end{cases} \qquad (6.1)$$

At the end of the iteration step, each node p_i is replaced by its new position

p_i'. This iteration step is repeated until either the maximal node movement

$$d_{\max} := \max_{i \in \{1,\dots,n_p\}} \|p_i - p_i'\|$$

drops below a prescribed threshold, or a prescribed maximal number of iteration steps is reached.

The regularizing polygon transformation given by Definition 5.1 is very well suited in order to compute $p_{i,j}'$, since the computation scheme only incorporates both the neighboring nodes of p_i with respect to the polygon E_j. With this, the parallelization of the computation of all p_i' according to (6.1) is straightforward and does not require any locking.

In contrast, the hexahedral element transformation scheme given by Definition 5.9 is comparably expensive, since it involves the computation of the dual element. In this case it is more efficient to compute the transformed element once, and add the resulting transformed element nodes $p_{i,j}'$ iteratively to the new node positions p_i', which are initialized as zero vectors. After the transformation of all elements, p_i' is divided by $|J(i)|$ to derive the arithmetic mean. This approach is also well suited for parallelization. However, it requires the updates of p_i' to be atomic. Since this element oriented version of basic GETMe simultaneous smoothing is the more general approach with respect to element transformation, it is summarized in Algorithm 1.

Algorithm 1: Basic GETMe simultaneous smoothing.

Input   : Mesh with nodes p_i, $i \in I$, and elements E_j, $j \in J$
Output : Smoothed mesh

1 **for** *iter* := 1 **to** maxIteration **do**
2   Set all p_i', $i \in I$, to zero vectors;
3   **foreach** *mesh element E_j* **do**
4          $\{p_{i,j}' : i \in I(j)\} \leftarrow$ `transformAndScaleElement`(E_j);
5   **foreach** $i \in I(j)$ **do**
6              $p_i' := p_i' + p_{i,j}'$;
7   **end**
8   **end**
9      $d_{\max} := 0$;
10   **foreach** *non-fixed node p_i* **do**
11         $p_i' := (1/|J(i)|)p_i'$;
12         $d_{\max} := \max(d_{\max}, \|p_i - p_i'\|)$;
13         $p_i := p_i'$;
14   **end**
15   **if** $d_{\max} \leq$ distanceThreshold **then**
16   **break**;
17   **end**
18 **end**

The for loop of line 1 represents the smoothing iteration loop. Within each iteration, all elements are transformed and all non-fixed nodes are updated. To do this, the new nodes p_i' are first initialized to zero vectors in order to compute the arithmetic means by iteratively adding the transformed element nodes $p_{i,j}'$.

Line 3 starts the transformation loop over each element of the mesh. In this, each element E_j is transformed and scaled, resulting in $|I(j)|$ new temporary nodes $p_{i,j}'$, which are added to the associated nodes p_i'. Here, the edge length sum-preserving scaling scheme is preferably applied. If the transformation does not preserve the element centroid, this function also translates the element to do so. Relaxation could also be applied but is omitted here for the sake of simplicity. The subsequent loop starting in line 10 computes the new positions of the non-fixed nodes by dividing the sum of the transformed temporary nodes by the number $|J(i)|$ of incident elements. In addition, the maximal node movement $d_{\max}$ is determined. After that, the node positions p_i are updated by p_i'.

Finally, smoothing is preliminarily terminated, if the maximal node movement drops below a given threshold. Here, this node movement-based termination criterion has been chosen to result in a simple smoothing scheme not incorporating any quality criteria, which makes it well suited for smoothing invalid meshes. Nevertheless, quality-based termination criteria could also be used if preferred.

The effect of the basic GETMe simultaneous smoothing approach, also denoted as *basic GETMe*, is illustrated exemplarily for the Platonic mesh depicted in Figure 6.2a, consisting of 54 regular quadrilaterals, 36 regular hexagons, and 23 regular dodecagons. The width and height of the mesh amount to 22.66 and 20.12, respectively. This mesh has been distorted by applying local random node movements to the inner mesh nodes, which invalidated 99 of the 113 elements. The resulting mesh depicted in Figure 6.2b has been used as initial mesh for the basic GETMe simultaneous approach. Smoothing was terminated, using a maximal node movement threshold of 0.01. Smoothing parameters have been set to $\lambda = 1/2$ and $\theta = \pi/n$ for n-gons. According to Lemma 5.7 this specific choice of θ represents the middle of the interval (θ_0, θ_1) for which the generalized polygon transformation has a regularizing effect.

For comparison reasons, the initial mesh has also been smoothed using the basic Laplacian smoothing approach based on neighboring node averaging using the same geometric termination criterion. In both cases, boundary nodes have been kept fixed during smoothing. The global optimization approach could not be applied due to the invalidity of the initial mesh.

Figures 6.2c and 6.2d depict the meshes obtained after applying ten iterations of basic Laplacian smoothing and basic GETMe smoothing, respectively. As in the case of the initial mesh, each element is shaded according to its mean ratio quality number, with dark gray indicating invalid elements and light gray indicating regular elements. The associated quality grayscale bar is shown below the meshes.

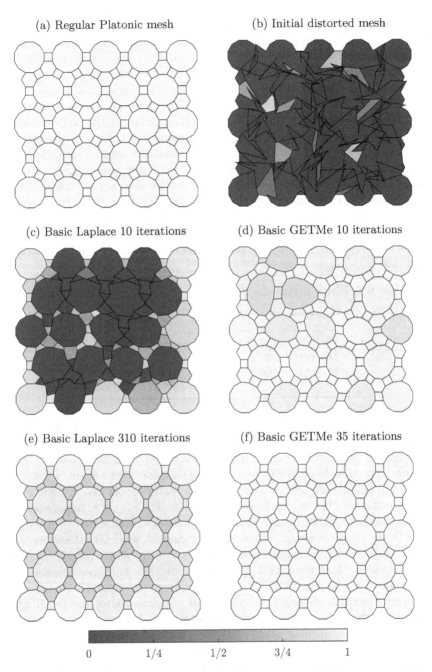

Figure 6.2: Platonic mesh smoothing example with mesh elements shaded according to their mean ratio quality number.

As can be seen, both methods lead to a significant mesh improvement within the first few steps. However, after applying 10 iterations of basic Laplacian smoothing, the mesh still contains 49 invalid elements. In contrast, all mesh elements are valid after applying the same number of iterations of basic GETMe smoothing. Furthermore, in contrast to the basic GETMe smoothing approach, basic Laplacian smoothing results in a non-regular configuration, as can be seen by comparing the final meshes depicted in Figures 6.2e and 6.2f.

This is also visible from Table 6.1, which provides the minimal mean ratio quality numbers for each element type and mesh. Due to the platonic configuration of the undistorted mesh, all elements are regular, resulting in the ideal value of one for all element types. After mesh distortion, invalid elements exist for all three element types as is indicted by the entry "invalid" in the table. Basic GETMe smoothing yields an almost regular mesh. Here, quality numbers slightly below one are due to the geometric termination criterion based on maximal node movements. In contrast, basic Laplacian smoothing results in q_{min} numbers for quadrilaterals and hexagons, which are significantly below one.

Table 6.1: Element type dependent q_{min} for Platonic mesh example.

	4-gons	6-gons	12-gons
Platonic mesh	1.0000	1.0000	1.0000
Distorted initial	invalid	invalid	invalid
Basic Laplace	0.8613	0.7482	0.9499
Basic GETMe	0.9971	0.9984	0.9955

This is due to the fact that basic Laplacian smoothing does not incorporate concepts for element regularization. Hence its node averaging approach leads to a different symmetric configuration, which depends on local mesh topology. Compared to this, basic GETMe simultaneous smoothing naturally results in the ideal regular configuration due to the incorporation of element-regularizing transformations. This driving force behind basic GETMe simultaneous smoothing also leads to a significant reduction of iteration numbers if compared to basic Laplacian smoothing. Whereas in the case of basic GETMe smoothing the maximal node movement drops below 0.01 after as few as 35 iterations, Laplacian smoothing requires 310 iterations, which is 8.9 times more.

Although neither basic GETMe simultaneous smoothing nor basic Laplacian smoothing incorporate any quality measures for smoothing and termination control, mesh quality is significantly improved, as is depicted in Figure 6.3. Here, the minimal element quality q_{min} and the mean mesh quality q_{mean} are depicted with respect to the iteration number.

In the case of basic GETMe simultaneous smoothing, all elements are valid after eight iterations, whereas basic Laplacian smoothing requires 58 iterations for the mesh to become valid. The non-regular result of the latter method is also visible from the final quality numbers, resulting in $q_{min} = 0.7482$ and

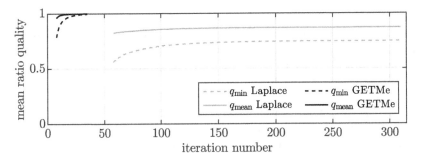

Figure 6.3: Minimal element and mean mesh quality with respect to smoothing iteration numbers.

$q_{mean} = 0.8701$ after 310 iterations. In contrast, the resulting quality numbers of the basic GETMe simultaneous approach after 35 iterations, amounting to $q_{min} = 0.9955$ and $q_{mean} = 0.9994$, are almost ideal. Both methods have in common that the first iterations in particular are very effective. After that, mesh quality improvement is comparably low. In the case of basic Laplacian smoothing, there are more than 200 iterations without a significant quality improvement until the maximal node movement drops below the prescribed threshold of 0.01.

6.1.2 GETMe simultaneous smoothing

Like every other geometry-based approach, basic GETMe simultaneous smoothing cannot guarantee mesh validity after every iteration step. Particularly in the case of unfavorable topological configurations, geometry-based smoothing may lead to the generation of invalid elements. Such configurations include the number of incident elements being too high or low, many fixed boundary nodes due to suboptimal discretization width, and polygonal elements with an increased number of nodes. Depending on the type of application, like the finite element method, such invalid elements are not acceptable. Furthermore, mesh quality has an impact on finite element solution accuracy. This is the reason why mesh smoothing should not only result in valid meshes, but also in high quality meshes. Here, mesh validity and quality is usually a matter of the same criterion.

In the following, the mean ratio quality criterion defined in Section 2.3.2 is used to present a quality controlled smoothing scheme. Here, the mean ratio quality criterion has been chosen for several reasons. First, it ensures element validity for finite element computations, since the Hessian of elements must be positive. Furthermore, it is defined for all element types under consideration including planar and non-planar polygons as well as the common polyhedral elements. This implies that all types of meshes can be smoothed using the same quality criterion and control mechanisms, which is favorable from an implementation point of view. Furthermore, mean ratio quality numbers are normalized to $[0, 1]$ with zero indicating invalid elements and one indicating regular elements.

This is also the reason why the mean ratio quality criterion is an excellent choice for local or global optimization approaches for mesh quality improvement. Thus, results of the quality controlled GETMe approach presented in the following can be compared with a state-of-the-art global optimization approach. Nevertheless, the presented smoothing approach can easily be adapted so as to use alternative quality criteria, as has been demonstrated in [160], combining the mean ratio quality criterion with the warpage criterion for quadrilateral element faces.

The quality-controlled GETMe simultaneous approach, denoted as *GETMe simultaneous*, presented in the following, is an advanced version of the basic GETMe simultaneous algorithm presented in the previous section. It is enhanced by adding quality controlled technics for the choice of transformation parameters, node averaging, and termination control. In addition, the generation of invalid elements is prevented by a simple technique of node resetting. Each of these modifications is now described in detail. As a prerequisite, the initial mesh to be smoothed must be valid.

The idea behind a quality dependent choice of mesh elements is to regularize low quality elements less aggressively, since they are usually situated in a local configuration where only minor changes in element shape can be applied in order not to invalidate neighboring elements. A natural choice to control the regularizing effect is the transformation parameter. For example, as has been shown in Figure 5.14 for the regularizing polygon transformation using $\lambda = 1/2$, the rate of regularization decreases with an increasing transformation angle θ. Similarly, the regularizing effect of the dual element-based transformation scheme for polyhedral elements decreases with an increasing scaling parameter $\sigma > 1/2$ as is depicted Figure 5.24. Hence, for polyhedral elements, an element type and element quality-related choice of the transformation parameters is given by

$$\sigma_j := \sigma_{\min}^{\text{type}} + \left(\sigma_{\max}^{\text{type}} - \sigma_{\min}^{\text{type}}\right)\left(1 - q(E_j)\right). \tag{6.2}$$

For type $\in \{\text{tet}, \text{hex}, \text{pyr}, \text{pri}\}$ the normals scaling factor σ is chosen depending on the actual element quality $q(E_j)$ within the element type-specific interval $[\sigma_{\min}^{\text{type}}, \sigma_{\max}^{\text{type}}]$. The larger $q(E_j)$, the lower σ_j, thus resulting in a more aggressive choice of the transformation parameter. In contrast, low quality elements are transformed using more conservative parameters near $\sigma_{\max}^{\text{type}}$. The same approach can be applied for choosing θ in the case of polygonal elements.

Without further modifications, the regularization of good quality elements using more aggressive transformation parameters with neighboring low quality elements using more conservative transformation parameters would prefer the already good elements. Hence, the transformed node averaging scheme is also modified to incorporate mesh quality aspects. The arithmetic mean is generalized to a weighted average according to

$$p_i' := \begin{cases} p_i & \text{if } p_i \text{ is fixed or } \tilde{w}_i = 0 \\ \frac{1}{\tilde{w}_i} \sum_{j \in J(i)} w_j p_{i,j}' & \text{otherwise,} \end{cases} \tag{6.3}$$

with

$$\tilde{w}_i := \sum_{j \in J(i)} w_j \quad \text{and} \quad w_j := \left(1 - q(E_j)\right)^\eta. \tag{6.4}$$

Again, for each non-fixed node p_i and its incident elements E_j, $j \in J(i) \subseteq J$, the new position p_i' is computed using the associated transformed node $p_{i,j}'$ per incident element. Each of these nodes is weighted by w_j. This choice of weights decreases the influence of high quality elements, whereas low quality elements are taken more into consideration. A special case occurs for a node with all incident elements being regular, which implies that the denominator $\tilde{w}_i$ becomes zero. In this special case, p_i' is not modified, i.e., $p_i' := p_i$, since the local configuration is already ideal.

The additional weight exponent $\eta \in \mathbb{R}_0^+$ in the definition of w_j according to (6.4) allows a non-linear control over the influence of high and low weights. In addition, the specific choice $\eta = 0$ results in the arithmetic mean, thus resulting in the basic GETMe smoothing scheme being a special case of GETMe smoothing that uses $\sigma_{\min}^{\text{type}} = \sigma_{\max}^{\text{type}}$ for each element type.

Due to the geometry-driven approach, the new node computation according to (6.3) cannot guarantee all mesh elements to be valid. Such elements must be identified and removed. For reasons of simplification, this is done by resetting the associated nodes of invalid elements to their previous position p_i. This again might invalidate neighboring elements. Hence this step must be applied iteratively until all elements are valid. However, usually invalid elements only occur in an early stage of mesh smoothing for regions of poor mesh quality. This is also the reason why more elaborate invalid element prevention techniques, such as applying node relaxation, would only add additional complexity to the algorithm without changing the result significantly.

Following a quality-based approach, it is also natural to use a quality-based termination control scheme instead of the geometric node movement criterion of the basic GETMe simultaneous smoothing approach. Thus, termination is based on the number of iterations performed so far and the mesh quality numbers of the current and previous iteration. This results in the GETMe simultaneous smoothing scheme given by Algorithm 2.

As in the case of the basic GETMe simultaneous smoothing algorithm, one iteration step of the smoothing process starts with initializing all new node positions p_i' to zero vectors in line 2. In addition, the variables $\tilde{w}_i$ are likewise initialized to zero. They will be updated during the element transformation loop to finally result in the node-related weight sums $\tilde{w}_i$ according to (6.4). In addition, the current mean mesh quality is stored in $q_{\text{mean\_old}}$, which will be used by the termination criterion at the end of the iteration.

The element transformation loop starting in line 4 first calls the function

---

Algorithm 2: GETMe simultaneous smoothing.

Input   : Valid mesh with nodes p_i, $i \in I$, and elements E_j, $j \in J$
Output: Smoothed valid mesh

1 **for** $iter := 1$ **to** maxIteration **do**
2   Set all p_i' to zero vectors and $\tilde{w}_i := 0$, where $i \in I$;
3      $q_{\text{mean\_old}} := q_{\text{mean}}$;
4   **foreach** *mesh element* E_j **do**
5          $\{p_{i,j}' : i \in I(j)\} \leftarrow$
                qualityTransformScaleAndRelaxElement(E_j);
6          $w_j := (1 - q(E_j))^\eta$;
7   **foreach** $i \in I(j)$ **do**
8              $p_i' := p_i' + w_j p_{i,j}'$;
9              $\tilde{w}_i := \tilde{w}_i + w_j$;
10   **end**
11   **end**
12   **foreach** *non-fixed node* p_i *with weight sum* $\tilde{w}_i > 0$ **do**
13         $p_i := (1/\tilde{w}_i)p_i'$;
14   **end**
15   iterativelyResetNodesOfInvalidElements();
16   **if** $|q_{\text{mean}} - q_{\text{mean\_old}}| \leq$ qualityThreshold **then**
17   **break**;
18   **end**
19 **end**

---

qualityTransformScaleAndRelaxElement for each element. This function uses quality-dependent transformation parameters like those given by (6.2) and applies the transformation to E_j. After that, the transformed element is scaled, the centroid is reconstructed if moved by the transformation, and relaxation is conducted according to Definition 5.6 and Definition 5.11, respectively, using a prescribed relaxation parameter ϱ. The resulting intermediate nodes $p_{i,j}'$ as well as the quality-based element weights w_j are then added to each successively updated sum of weighted nodes p_i' and node-based weight sums $\tilde{w}_i$ in the loop starting from line 7.

New node computation and setting is conducted in the loop starting from line 12. It consists of computing the weighted averages and setting the resulting new node positions if the node is non-fixed and the weight sum $\tilde{w}_i$ is greater than zero.

As mentioned before, setting new node positions might lead to invalid elements. Hence, the iterative element resetting scheme iterativelyReset-NodesOfInvalidElements is applied. In this all nodes of invalid elements are determined. Their coordinates are set back to the values p_i before their modification in the node update loop starting from line 7. If invalid elements are generated again, this step is repeated until no further invalid element exists. Termination of this iterative resetting scheme is guaranteed, since the mesh is valid in the beginning of each smoothing iteration.

The iteration is repeated until either a given maximal number of iterations is reached or the mean mesh quality improvement drops below a given quality threshold. The latter is checked in line 16 with the aid of the previous iteration mean mesh quality $q_{\text{mean\_old}}$. Here, the mean mesh quality is used, since the simultaneous smoothing approach is based on transforming all elements within one iteration step, resulting in the mean mesh quality being the appropriate measure of the mesh modification. Instead of a preliminary termination after the first iteration without mean mesh quality improvement, a more conservative approach would be to terminate the smoothing process after a given number of consecutive smoothing iterations without significant q_{mean} improvement.

6.1.3 GETMe sequential smoothing

By transforming all elements and determining new node positions as weighted averages, the simultaneous GETMe smoothing approach presented in the previous section is focused on effectively improving overall mesh quality. That is, on improving the arithmetic mean q_{mean} of all element quality numbers. However, for example, in the case of the finite element method, single low quality elements can deteriorate solution efficiency and the resulting accuracy. Hence, in this case, improving the lowest mesh element quality q_{min} is also of particular interest.

If boundary nodes are kept fixed during the smoothing process, the element with the lowest quality might consist entirely of boundary nodes. Since in this case q_{min} cannot be improved, q_{min}^* denoting the minimal element quality of all elements with at least one interior node will be used instead. This number gives the lowest quality of all elements with a non-fixed quality number, hence elements with improvement potential.

Improving this quality number is specifically addressed by the *GETMe sequential smoothing* approach presented in this section. It is based on the simple idea that a single element can be improved by applying a regularizing transformation, resulting in a slightly modified element with better quality. Setting new node positions for this element also affects neighbor elements. However, using a proper choice of transformation and relaxation parameters, this results in new local configuration with better quality.

This approach is illustrated by Figure 6.4, depicting on the left side a mesh with twelve triangular elements. For each element, the mean ratio quality number is given. As can be seen, the lowest quality element E with $q(E) = 0.56$ is located in the mesh center. It is transformed using the regularizing polygon transformation (5.4) with $\lambda = 1/2$ and $\theta = \pi/3$. In addition, edge length-preserving scaling according to Definition 5.5 and relaxation according to Definition 5.6 using $\varrho = 1/2$ has been applied. This results in the better-shaped triangle marked gray on the left of Figure 6.4. The mesh depicted on the

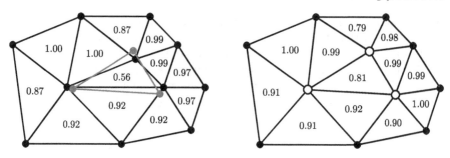

Figure 6.4: Lowest quality element transformation by GETMe sequential.

right is obtained by setting the nodes of the lowest quality element to the positions of the transformed element marked by circles. In doing so, the mean ratio quality number of the center element is significantly improved to 0.81. In addition, neighboring element quality numbers slightly change, since they are also affected by these node modifications.

Following this minimally invasive approach, GETMe sequential smoothing is based on iteratively picking the element with the lowest quality, transforming it, and setting new node positions directly in the mesh. Here the word sequential in the name of this smoothing scheme indicates that only one element is transformed per iteration step.

In the case of adverse local topological configurations, the element with the lowest quality might not be further improvable, or, transforming this element might even invalidate neighboring elements. In such situations, this simple approach gets stuck by always picking the same element, which is not further improvable. Hence, the algorithm is refined by a quality penalty approach adding additional control over the element selection mechanism. For each element E_j, $j \in J$, with quality $q(E_j)$, a quality penalty term $\pi_j \geq 0$ is introduced, resulting in the corrected element quality $q(E_j) + \pi_j$. This results in Algorithm 3, for which details will be discussed in the following.

First, all element quality penalty parameters π_j are initialized to zero in line 1 and the index j_{old} of the last selected element is set to -1. After this, the main smoothing loop is applied with a predefined maximal number of iterations. In contrast to the GETMe simultaneous approach, which transforms all mesh elements at once within one iteration, the GETMe sequential approach transforms only one element per iteration.

Within one smoothing iteration step, the index of the element with the lowest penalty-corrected quality number $q(E_j) + \pi_j$ is selected first in line 4. After that, this element is transformed, scaled, and relaxed, resulting in the new intermediate nodes $p'_{i,j}$. In this context, scaling is particularly sensible, since the element growth caused by the transformation might affect neighboring elements too much. In addition, relaxation is applied according to Definition 5.6 or Definition 5.11 for polygonal and polyhedral elements, respectively. The latter allows a better control of the rate of regularization, which is required

---

Algorithm 3: GETMe sequential smoothing.

Input : Valid mesh with nodes p_i, $i \in I$, and elements E_j, $j \in J$
Output : Smoothed valid mesh

1 Set element quality penalty values $\pi_j := 0$ for all $j \in J$;
2 $j_{\text{old}} := -1$;
3 **for** $iter := 1$ **to** maxIteration **do**
4   Find $j \in J$ with $q(E_j) + \pi_j \leq q(E_k) + \pi_k$ for all $k \in J$;
5     $\{p'_{i,j} : i \in I(j)\} \leftarrow$ transformScaleAndRelaxElement(E_j);
6   **foreach** $i \in I(j)$ *with* p_i *not fixed* **do**
7         $p_i := p'_{i,j}$;
8   **end**
9   **if** *elements have been invalidated* **then**
10   Reset all p_i, $i \in I(j)$, to their previous position;
11         $\pi_j := \pi_j + \Delta\pi_i$;
12   **else**
13         $\pi_j := \max(0, \pi_j - \Delta\pi_s)$;
14   **end**
15   **if** $j = j_{\text{old}}$ **then**
16         $\pi_j := \pi_j + \Delta\pi_r$;
17   **end**
18     $j_{\text{old}} := j$;
19   **if** *number of consecutive iterations without* $q^*_{\min}$ *improvement* $>$ iterThreshold **then**
20   **break**;
21   **end**
22 **end**

---

in the case of problematic local configurations, where moderate geometric changes are required in order not to invalidate neighboring elements. All these steps result in the new element node coordinates $p'_{i,j}$, which are set in line 7 of Algorithm 3.

Setting these transformed nodes might invalidate direct neighboring elements. Hence, an invalid element-handling step is required as in the case of the GETMe simultaneous approach, to guarantee mesh validity. Due to good experimental results and in order not to increase algorithmic and computational complexity, a simple resetting approach was chosen. That is, in case invalid elements occur, the nodes of the transformed element are simply reset to their previous untransformed positions. Without further modifications, the next iterations would select, transform, and reset the same element repeatedly without any quality improvements.

This is prevented by adjusting element quality by a simple penalty mechanism. If invalid elements have been generated, the penalty term π_j of the selected element E_j is increased by a fixed invalid element penalty value

$\Delta \pi_i > 0$. Since low quality elements tend to occur clustered within a mesh, this results in other elements being selected and improved first, until the postponed element can be transformed without invalidating its neighbors. For the same reason, transforming the same element repeatedly too often should be avoided. Therefore, if the element is selected repeatedly, an additional penalty term $\Delta \pi_r > 0$ is added. In contrast, if the element has been successfully transformed, the element penalty value π_j is decreased by the success penalty value $\Delta \pi_s > 0$ if the result is not a negative number. This invalid element and penalty handling is conducted from line 9 to line 18.

Smoothing is terminated if either the maximal number of iterations is reached, or the number of consecutive iterations without improvement is larger than a given threshold. If the mean mesh quality is also of interest, one can additionally terminate if the deterioration of q_{mean} caused by the GETMe sequential approach becomes too large. However, since the computation of q_{mean} is computationally expensive, such a criterion should only be checked each nth iteration, with n chosen sufficiently large.

It is recommended that the lowest quality element selection in line 4 be based on using a min heap of the penalty-corrected quality numbers and associated element indices. Determining the element index j using such a min heap is an $\mathcal{O}(1)$ operation in contrast to the $\mathcal{O}(n_e)$ operation of finding the minimum of all n_e element quality numbers. After setting new nodes, the quality values of all incident elements must be updated in the min heap.

6.2 Combined GETMe algorithms

6.2.1 GETMe smoothing

By deriving new node positions as weighted means of transformed element nodes for the entire mesh, the GETMe simultaneous approach focuses on improving the overall mesh quality q_{mean}. In contrast, transforming only low quality elements in the case of GETMe sequential puts an emphasis on improving the lowest element quality number q_{min}^*. These individual strengths with respect to quality and runtime behavior can be combined by applying GETMe simultaneous smoothing first, which improves overall mesh quality and with these problematic parts of the mesh containing the element with the lowest quality. Subsequently applying GETMe sequential smoothing can then improve the remaining low quality elements more effectively. The resulting combined approach is denoted as *GETMe smoothing* and is summarized for completeness in Algorithm 4.

Due to its approach of smoothing all mesh elements, hence modifying all non-fixed mesh nodes, it is crucial to apply GETMe simultaneous before applying GETMe sequential. Otherwise, GETMe simultaneous would override

---

Algorithm 4: GETMe smoothing.

Input : Valid mesh
Output : Smoothed valid mesh

1 Apply GETMe simultaneous smoothing;
2 Apply GETMe sequential smoothing;

---

the results obtained by GETMe sequential. Since the latter usually only modifies a small subset of the mesh, both methods are good to combine.

6.2.2 Adaptive GETMe smoothing

The two-stage approach of GETMe smoothing, being based on improving all mesh elements by GETMe simultaneous in the first substep, and single lowest quality element improvement by GETMe sequential in the second substep, is easy to implement and leads to high quality meshes. However, this approach does not specifically address meshes comprising both high quality submeshes requiring almost no improvement and low quality submeshes requiring significant improvement. Such meshes are generated, for example, by mesh generation techniques filling the interior of the domain by a regular or at least almost regular template mesh. Here, individually shaped elements are only generated for interface layers attaching the interior mesh to the boundary of the domain. On the other hand, improving the lowest quality element often requires the modification of its neighboring elements as well, to improve local mesh quality.

Therefore, it is an obvious approach to find a compromise between the two extremes of transforming either all mesh elements or only one. This novel approach, first presented in [153], is based on smoothing submeshes, which are identified by element quality. From an algorithmic point of view, smoothing control is simplified, the number of parameters is reduced, if compared to GETMe smoothing, and parallelization is improved. In addition, by using an adaptive node relaxation technique instead of invalid element node resetting, an additional flexibility is added to handle problematic mesh regions. This results in the GETMe adaptive smoothing approach given in Algorithm 5.

A quality threshold q_t is first initialized with a prescribed value. It is used to identify those elements that should be improved first. Without further knowledge of the mesh, this parameter can be initialized to $q_t := 1$, resulting in a behavior similar to that of GETMe simultaneous smoothing, which transforms all mesh elements in the first stage. Therefore, the first cycle is denoted as mean cycle running, for which the current algorithm state is initialized in line 2. After this, the node relaxation values vector $R := (\varrho_1, \dots, \varrho_k)$ is initialized for this specific state. The role of R will be described later in the context of the iterative node relaxation scheme.

The smoothing loop of GETMe adaptive starting from line 4 comprises

---

Algorithm 5: GETMe adaptive smoothing.

Input : Valid mesh
Output: Smoothed valid mesh

1 $q_t \leftarrow$ getInitialQualityThreshold();
2 state := MeanCycleRunning;
3 $R \leftarrow$ getNodeRelaxationValuesVector(*state*);
4 **for** *iter* := 1 **to** maxIteration **do**
5      $p'_i \leftarrow$ computeNewNodes(q_t, p_i, E_j);
6   updateNodesByIterativeNodeRelaxation(R, p_i, p'_i);
7   **if** *state* = MeanCycleRunning *and*
       $\Delta q_{\mathrm{mean}} \leq$ qmeanImprovementThreshold **then**
8   state:= MinCycleStart;
9          $R \leftarrow$ getNodeRelaxationValuesVector(*state*);
10   **end**
11   **if** *state* = MinCycleRunning **then**
12   **if** *no q^*_{min} improvement in last iteration* **then**
13   noQMinImproveCounter := noQMinImproveCounter + 1;
14   **end**
15   **if** *noQMinImproveCounter* > maxNoQMinImproveCounter **then**
16   state := MinCycleStart;
17   **end**
18   **end**
19   **if** *state* = MinCycleStart **then**
20   **if** *no q^*_{min} improvement in last min cycle* **then**
21   **break**;
22   **end**
23   state := MinCycleRunning;
24   noQMinImproveCounter := 0;
25         $q_t \leftarrow$ determineTransformationThreshold();
26   **end**
27 **end**

---

the following steps. First, new node positions are computed by the substep computeNewNodes. After that, new node positions are set by an iterative node relaxation technique in updateNodesByIterativeNodeRelaxation, which also avoids the generation of invalid elements. Both substeps will be described in detail later in this section, in order to discuss smoothing-control aspects first. The latter are given by the lines 7 to 26 of the algorithm.

As noted before, smoothing is performed in two stages starting with the mean cycle. The termination of this cycle is controlled by the if statement in line 7. This mean cycle is applied as long as the mean mesh quality improvement Δq_{mean} of two consecutive smoothing iterations is larger than a given threshold. If the mean quality improvement drops below this threshold, the state variable is set to MinCycleStart, which indicates that the initialization of a second stage cycle must be conducted. Furthermore, the node relaxation values vector R is changed, since a more conservative relaxation strategy should be applied for the problematic regions of the mesh.

The second stage, which is geared towards improving q_{min}^*, is divided into individual smoothing cycles accomplishing an adaptive number of smoothing iterations each. Furthermore, each cycle is characterized by its individual element selection quality threshold q_t. The termination of such a cycle is controlled by the if statement starting in line 11. First, whether the minimal improvable element quality q_{min}^* has not been improved by the last iteration is checked. If this is the case, the noQMinImproveCounter is increased. A new minimal element quality improvement cycle is started if the noQMinImproveCounter exceeds a given number of iterations defined by maxNoQMinImproveCounter. This is done by setting the state variable to MinCycleStart.

The initialization of a new min cycle is handled by the content of the if-clause starting from line 19. First, whether q_{min}^* was improved by the last cycle is checked. If this is not the case, smoothing is preliminarily terminated. Otherwise, a new min cycle is initialized by setting the state variable to MinCycleRunning, by resetting the noQMinImproveCounter to zero, and by selecting a new element quality threshold q_t. Such a choice of q_t can be based on taking the quality number of a prescribed position in the sorted vector of all element quality numbers; for example, taking the largest quality value of the 10% of those mesh elements with the lowest element quality.

As can be seen, this simple control technique conducts the single smoothing cycle of the mean mesh quality-oriented first stage and the individual number of smoothing cycles for min mesh quality improvement in the second stage within one main loop. Here the two stages are motivated by switching the quality criterion from q_{mean}, used first, to q_{min}^* in the second stage. This approach could be reduced even further to one stage if the quality criterion used for termination and its associated quality threshold are both selected dynamically.

It remains to describe the two substeps `computeNewNodes` and `update-NodesByIterativeNodeRelaxation`, which are relevant for determining node positions after one smoothing loop. The steps for new node computation are given by Algorithm 6.

Independent of the smoothing stage, new nodes are computed as weighted averages of transformed elements. In the case of GETMe simultaneous smoothing, the weighted averaging scheme given by (6.4) is based on weighting intermediate nodes based on their associated element qualities. In the case of GETMe adaptive smoothing, the following alternative weighing scheme based on neighboring element qualities is used instead. It is computationally more expensive, but takes local submesh quality more into account. Here transformed nodes are also computed according to (6.3). However, the weights are given by

$$w_j := \sqrt{\frac{\sum_{n \in N(j)} q(E_n)}{|N(j)| q(E_j)}}, \tag{6.5}$$

Algorithm 6: Function `computeNewNodes` of GETMe adaptive.

Input : Quality threshold q_t, node positions p_i and Elements E_j

Output: New node positions p_i'

1 Set all p_i' to zero vectors and $\tilde{w}_i := 0$, where $i \in I$;

2 **foreach** *mesh element E_j* **do**

3      $w_j := \sqrt{\left(\sum_{n \in N(j)} q(E_n)\right)/\left(|N(j)|q(E_j)\right)}$;

4   **if** $q(E_j) \leq q_t$ **then**

5          $\{p_{i,j}' : i \in I(j)\} \leftarrow$ `transformAndScaleElement`(E_j);

6   **else**

7          $p_{i,j}' := p_i$ for $i \in I(j)$;

8   **end**

9   **foreach** $i \in I(j)$ **do**

10          $p_i' := p_i' + w_j p_{i,j}'$;

11          $\tilde{w}_i := \tilde{w}_i + w_j$;

12   **end**

13 **end**

14 **foreach** $i \in I$ **do**

15   **if** p_i *is non-fixed and* $\tilde{w}_i \neq 0$ **then**

16          $p_i' := (1/\tilde{w}_i)p_i'$;

17   **else**

18          $p_i' := p_i$;

19   **end**

20 **end**

---

with $N(j)$ denoting the index set of the neighbor elements of E_j, i.e., elements that share at least one common node with E_j. The number of such neighbor elements is denoted as $|N(j)|$.

This modified weight given by (6.5) is based on the quotient of the average neighbor element quality and the quality of E_j itself. Hence w_j is larger than one if the quality of E_j is below the average quality of its neighbors, and smaller than one if the quality of E_j is larger than the average quality of its neighbors. This puts an emphasis on weights of lower quality elements with respect to neighboring element quality.

Within the new nodes computation step given in line 1 of Algorithm 6, all values of p_i and $\tilde{w}_j$ are initialized to zero. This is followed by an element loop in which the nodes weight w_j for the element E_j is determined first. If the quality of the element E_j is below or equal to the quality threshold q_t, the element is transformed using fixed transformation parameters and scaled to preserve its initial edge length sum. Otherwise its current node positions p_i are used as intermediate node positions $p_{i,j}'$ in line 7.

In the subsequent loop starting from line 9, the weighted intermediate nodes and weights are summed up in the variables p_i' and $\tilde{w}_i$ to derive the final new nodes according to (6.3) using the weights given by (6.5). The resulting new node positions according to this equation are determined in the subsequent for loop for those nodes, which are non-fixed. Fixed nodes preserve their initial position.

The element quality-based selection of the elements to be transformed can also be enhanced by a local proximity aspect; for example, by also transforming the neighbors of low quality mesh elements. However, for simplicity reasons, such improvements are not considered here. It should also be noticed that in contrast to GETMe simultaneous smoothing, the choice of transformation parameters does not depend on element quality. This is due to the usage of an adaptive relaxation approach described next.

The iterative node relaxation step `updateNodesByIterativeNodeRelax-ation` called in line 6 of the GETMe adaptive Algorithm 5 replaces the iterative node resetting scheme of GETMe simultaneous smoothing to avoid the generation of invalid elements. Here, new nodes positions are not just reset if they lead to invalid elements, but iteratively adapted by relaxation. The definition of this function is given by Algorithm 7.

Algorithm 7: Function `updateNodesByIterativeNodeRelaxation` of GETMe adaptive.

Input   : Relaxation values vector R of dimension k, nodes p_i and p_i'
Output : Updates node positions p_i

1 $I_R :=$ set of all non-fixed node indices;
2 Set $p_i^{\text{old}} := p_i$ for each $i \in I_R$;
3 **for** $n := 1$ **to** k **do**
4   **foreach** $i \in I_R$ **do**
5          $p_i := (1 - \varrho_n)p_i^{\text{old}} + \varrho_n p_i'$;
6   **end**
7      $I_r :=$ set of all non-fixed node indices, with incident invalid
          elements;
8   **if** I_r *is empty* **then**
9   **break;**
10   **end**
11 **end**

First, the index set I_R of nodes, which must be updated by relaxation, is initialized by the index set of all non-fixed mesh nodes. Subsequently, line 2 saves for each non-fixed node the current node position p_i in the backup variable $p_i^{\text{old}} := p_i$.

In the subsequent relaxation loop starting from line 4, each node p_i with index $i \in I_R$ is updated by relaxation. That is, it is set to the linear combination of the initial position p_i^{old} of the current smoothing loop and the transformed position p_i' according to

$$p_i := (1 - \varrho_n)p_i^{\mathrm{old}} + \varrho_n p_i', \quad i \in I_R, \tag{6.6}$$

using a relaxation parameter ϱ_n. The latter is an element of the prescribed vector $R := (\varrho_1, \ldots, \varrho_k)$ of successively descending relaxation values.

As in the case of GETMe simultaneous smoothing, setting these new positions might result in the generation of invalid elements. Therefore, the index set I_r is set to the indices of all inner nodes that belong to invalid elements. But instead of resetting these nodes, a more conservative relaxation parameter is applied. This is done by using the next relaxation parameter r_{n+1} by the enclosing for loop starting from line 3.

By setting the last relaxation parameter to $\varrho_k := 0$, the validity of the resulting mesh is guaranteed, since for this case (6.6) simplifies to $p_i := p_i^{\mathrm{old}}$. That is, if relaxation cannot avoid the generation of invalid elements, the nodes are reset to their original valid position, like in the case of the GETMe approach.

A more aggressive smoothing strategy during the first mean stage of the algorithm can be based on applying overrelaxation, that is, by a choice $\varrho_1 > 1$. In contrast, for the choice $\varrho_1 := 1$, Equation (6.6) simplifies to $p_i := p_i'$. Hence, the transformed element nodes are used directly. The subsequent monotonically decreasing relaxation values are only used if invalid elements occur. Otherwise relaxation is preliminarily terminated by line 8. Aiming at low quality element improvements during the second stage of the algorithm, R is chosen more conservatively in line 9 of the GETMe adaptive scheme given in Algorithm 5.

From an algorithmic point of view, GETMe adaptive differs from GETMe smoothing in the following points [153]:

○ Incorporation of two smoothing stages within one smoothing loop instead of applying two separate loops.

○ Relaxation is applied to both previous and new node positions instead of involving previous and transformed elements. Furthermore, the relaxation parameter is adjusted on a nodal basis, which also replaces the iterative invalid element node-resetting scheme.

○ During the q_{min}^*-oriented stage, GETMe adaptive transforms more than one element per iteration, which is more suitable for parallel mesh smoothing. Furthermore, both stages use the same node averaging approach. However, nodes of elements E_j with $q(E_j) > q_t$ are directly used without transforming and scaling.

○ Iterations of the second stage are organized in cycles. The transformation quality threshold q_t is updated at the beginning of each smoothing cycle.

○ Fixed element transformation parameters are used instead of quality adaptive transformation parameters. The adjusted weights in the transformed element nodes averaging scheme are based on neighboring element quality ratios.

   ○ The exponent η of the node weights w_j given in Equation (6.4) and the penalty parameters of GETMe sequential are not used. This eliminates the need for a minimal element quality heap required by the sequential substep of GETMe.

For the efficient implementation of GETMe it is admissible to use parallelization. In addition, unnecessary element mean ratio quality evaluations should be avoided, since these computations are comparably expensive. That is, element quality is stored and only updated after element nodes have changed. This makes the neighbor quality-based node weight computation according (6.5) less expensive.

Furthermore, GETMe adaptive can be combined with the basic GETMe simultaneous untangling preprocessing step described in Section 6.4.4, which has two benefits. First, it makes GETMe adaptive applicable to tangled meshes. Second, the particularly effective and inexpensive basic GETMe simultaneous smoothing scheme is expected to speed up smoothing significantly.

6.3 Properties of GETMe smoothing

There are as yet no analytic proofs for the convergence and properties of the GETMe algorithms using the transformations presented in Chapter 5. However, this section will demonstrate some observed properties of GETMe smoothing on the example of simple model problems.

Results for a broad variety of real-world examples can be found in Chapter 7. Furthermore, a link to gradient flow-based optimization methods is established in Chapter 8, which gives an indication why GETMe smoothing results in a mesh quality similar to that obtained by applying global optimization.

6.3.1 Influence of transformation parameters

It has been shown in Section 5.1.2.2 that generalized polygon transformation has a regularizing effect if the base angle θ and the subdivision ratio λ are taken from the specific parameter domain D_1. The proof is based on identifying the dominant eigenvalue of the circulant transformation matrix, which also gives an indication of the rate of element convergence. In this section, the influence of element convergence on mesh smoothing convergence will be demonstrated with the example of a small quadrilateral mesh improved by basic GETMe simultaneous smoothing.

The initial quadrilateral mesh of width 100 and height 80 depicted in Figure 6.5a consists of 134 boundary nodes kept fixed during smoothing, 655 inner nodes, and 722 elements. As can be seen, the initial mesh is distorted by random inner node movements leading to 464 invalid elements. An improved

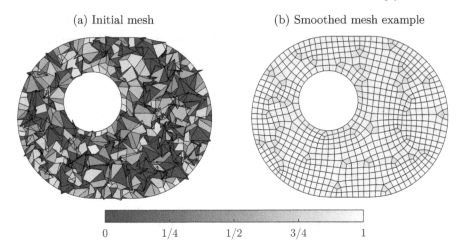

(a) Initial mesh   (b) Smoothed mesh example

$$0 \qquad 1/4 \qquad 1/2 \qquad 3/4 \qquad 1$$

Figure 6.5: Basic GETMe simultaneous parameter variation mesh example.

version of the same mesh with elements shaded according to their mean ratio quality number is depicted in Figure 6.5b. It has been smoothed by basic GETMe simultaneous smoothing according to Algorithm 1 using the generalized polygon transformation given by (5.6).

As stated by Theorem 5.1, this transformation has a regularizing effect for $(\theta, \lambda) \in D_1$ with the parameter domain D_1 depicted in Figure 5.9. Here, the rate of element convergence depends on the eigenvalue quotients, which are depicted in Figure 5.11 for the specific choice $\lambda = 1/2$. The smaller the ratio η_k/η_1 for $k \in \{2, 3\}$, the stronger the regularizing effect, which is the case for smaller values of θ. This imposes the question of which parameter combinations $(\theta, \lambda) \in D_1$ are particularly useful with respect to smoothing convergence and resulting mesh quality, which will be answered by the following numerical test.

The closed parameter interval $[0, 1]$ of λ has been discretized 401 equidistant values λ_k and the closed parameter interval $[0, \pi/2]$ of θ by 630 equidistant values θ_j. This leads to 252,630 elements (θ_j, λ_k) in the Cartesian product of both discretized parameter intervals. 125,498 of these parameter pairs are located inside the parameter domain D_1 resulting in regularizing quadrilateral transformations. For each of the latter parameter pairs, the initial mesh depicted in Figure 6.5a has been smoothed by basic GETMe simultaneous smoothing applying the generalized polygon transformation using (θ_j, λ_k) until the maximal node movement $d_{\max}$ dropped below 0.01. For each parameter combination, this results in the iteration number of basic GETMe simultaneous smoothing and the final mesh quality numbers $q_{\min}$ and $q_{\mean}$.

Figure 6.6 depicts the parameter domain D_1 shaded according to the resulting iteration number for each (θ_j, λ_k). The grayscale bar providing the associated number of iterations per shade of gray is depicted on the right.

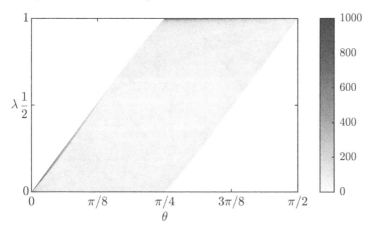

Figure 6.6: Resulting basic GETMe simultaneous iteration number per transformation parameter combination (θ_j, λ_k).

Here, low iteration numbers are marked light gray, large iteration numbers dark gray.

The achieved iteration numbers range from 44 to 921. However, as can be seen by the almost light gray area, most of the iteration numbers are very low. To be precise, 74.96% of the iteration numbers are less than or equal to 50, and 97.53% are less than or equal to 100. A close look at Figure 6.6 reveals that larger iteration numbers only occur close to the boundary of D_1. For values of λ near zero and one, this is due to the reduced regularizing effect, since the triangles constructed on the sides of each quadrilateral are almost degenerated to a straight line. For large values of θ, this is due to the reduced regularizing effect caused by large eigenvalue quotients. Nevertheless, the iteration number is consistently low inside the parameter domain D_1.

The parameter domain D_1 shaded by the resulting quality numbers of the smoothed mesh using the parameter pair (θ_j, λ_k) are depicted in Figure 6.7. Here, Figure 6.7a shows the lowest element quality number of the mesh, which ranges from $q_{\min} = 0$, indicating remaining invalid elements in the smoothed mesh, to the best achieved quality value $q_{\min} = 0.8610$. In 97.73% of all parameter pair cases, smoothing resulted in a valid mesh. The lowest minimal element quality $q_{\min}$ of all valid smoothed meshes amounts to 0.501 and in 83.48% of all cases $q_{\min}$ is larger than 0.75.

As can be seen in Figure 6.7a by the change of gray shades along a horizontal line, for example, defined by the choice $\lambda = 1/2$, the minimal element quality of the smoothed mesh decreases if θ increases. This is in accordance with the eigenvalue quotients depicted in Figure 5.11. Here small quotients η_k/η_1 for $k \in \{2, 3\}$ can be observed for small values of θ leading to a strong regularizing effect of the related transformation. The regularizing effect decreases if θ becomes larger. With this, the minimal element quality decreases too. That

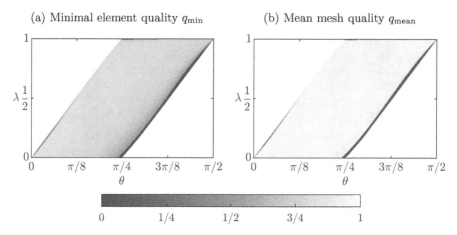

Figure 6.7: Resulting mesh qualities per transformation parameter combination (θ_j, λ_k) used by basic GETMe simultaneous smoothing.

is, the minimal element quality number of the smoothed mesh reflects the eigenvalue ratio and with this the rate of element convergence.

Figure 6.7b depicts the parameter domain D_1 shaded according to the resulting mean mesh quality number. Here, q_{mean} has been set to 0 if there is at least one invalid element in the mesh. In the case of the resulting 122,655 valid meshes, q_{mean} ranges from 0.9581 to 0.9757. That is, the choice of transformation parameters has only minor impact on the resulting mesh quality q_{mean}. This also implies that in all resulting meshes improved by basic GETMe simultaneous smoothing, the number of low quality or invalid elements is very low.

For the given example one can conclude that with respect to the resulting iteration number and mean mesh quality, basic GETMe smoothing is highly stable inside D_1. With respect to minimal element quality, using lower values of θ is recommended. In both cases and due to symmetry reasons of the resulting geometric transformations, $\lambda = 1/2$ is a suitable choice.

6.3.2   Influence of initial mesh validity and quality

The previous section indicated that basic GETMe simultaneous is quite stable with respect to the transformation parameters. In this section, the influence of initial mesh quality on smoothed mesh quality and the number of smoothing iterations is analyzed. This is done by smoothing various distorted versions of the mesh that was also considered in the previous section. The initial undistorted mesh is of width 100 and height 80. Starting from this, each inner node has been dislocated in a node-individual random direction with a node-individual random distance less than or equal to a prescribed maximal absolute

(a) Basic GETMe simultaneous iteration numbers

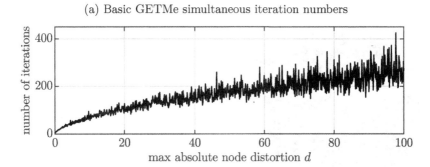

(b) Resulting quality numbers for smoothed meshes

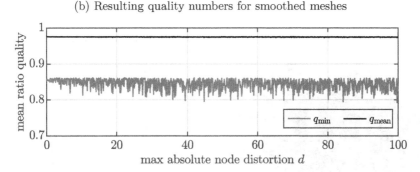

Figure 6.8: Basic GETMe simultaneous smoothing of increasingly distorted meshes.

node distortion distance $d \in [0, 100]$. Here $d = 0$ represents the undistorted initial mesh with $q_{\min} = 0.8517$ and $q_{\mathrm{mean}} = 0.9757$.

The interval $[0, 100]$ has been discretized into 1,000 equidistant values. For each of these values, a distorted mesh has been generated with d set to this specific value. After this, each distorted mesh has been smoothed using basic GETMe simultaneous smoothing and the fixed transformation parameters $\lambda = 1/2$ and $\theta = 0.4345$ in all cases. As in the case of the previous section, smoothing was terminated as soon as the maximal node movement dropped below 0.01.

Figure 6.8a depicts the resulting basic GETMe simultaneous iteration numbers as a function over the maximal absolute node distortion d. As can be seen, on average the iteration number increases if the distortion increases. The minimal and maximal iteration numbers amount to 3 and 425, respectively.

In contrast to the iteration numbers, the resulting mesh quality numbers do not vary to the same extent, as can be seen in Figure 6.8b. In all cases, $q_{\min} \in [0.7921, 0.8620]$ and $q_{\mathrm{mean}} \in [0.9735, 0.9759]$ hold. That is, the resulting mesh quality is quite stable with respect to $q_{\min}$ and particularly stable with respect to q_{mean}. Hence, the given example indicates that the basic GETMe

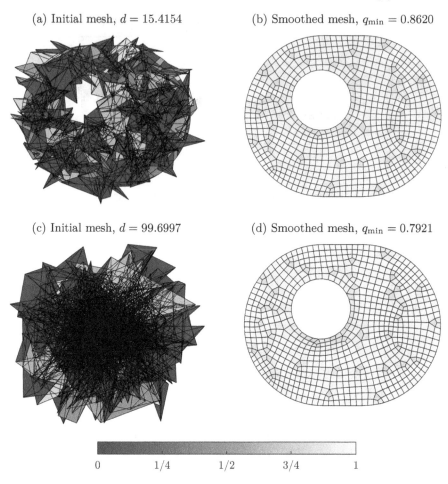

(a) Initial mesh, $d = 15.4154$ (b) Smoothed mesh, $q_{\min} = 0.8620$

(c) Initial mesh, $d = 99.6997$ (d) Smoothed mesh, $q_{\min} = 0.7921$

Figure 6.9: Initial and smoothed meshes with worst and best resulting minimal element quality.

smoothing result is not particularly sensitive with respect to the quality of the input mesh.

This is also illustrated by Figure 6.9. Here, Figure 6.9a depicts the initial mesh for the choice $d = 15.4154$, which resulted in 610 invalid elements. The associated smoothed mesh obtained after 84 iterations is depicted in Figure 6.9b. It is also the mesh with the best minimal element quality achieved for all cases, amounting to $q_{\min} = 0.8620$.

In contrast, Figure 6.9d depicts the smoothed mesh with the lowest minimal element quality of all cases, amounting to $q_{\min} = 0.7921$. This mesh has been obtained by smoothing the mesh depicted in Figure 6.9c, which has been generated by setting $d = 99.6997$. As can be seen, due to this extreme mesh distortion caused by the large value of d, the structure of the mesh and the domain are not visible. In particular, 636 of the 722 initial mesh elements are invalid. Nevertheless, basic GETMe simultaneous smoothing yields a high quality mesh within 253 iterations with only slight differences if compared to the mesh with the best q_{min} value, depicted in Figure 6.9b.

6.3.3 Unrestricted surface mesh smoothing

One of the observed features of basic GETMe simultaneous smoothing is its tendency to preserve locality during unrestricted polygonal surface mesh smoothing. Here, "unrestricted" means that no measures are considered to preserve the initial shape of the entire mesh, hence the model, during smoothing. That is, shape preservation techniques are not incorporated into the smoothing process. Preserving locality is a favorable behavior, since it supports locally acting techniques to preserve the local shape of the model and to ensure element validity. Therefore, compensation techniques to avoid global node movements are not required.

The tendency to preserve locality is illustrated by smoothing the two surface mesh examples depicted in Figure 6.10. One is a quadrilateral surface mesh of a flange model, shown on the left, the other a triangular surface mesh of a stub axle, depicted on the right. The STEP file of the underlying geometry model of the stub axle is provided courtesy of INPG by the AIM@SHAPE Mesh Repository [1].

Both initial meshes have been smoothed by applying the basic Laplacian smoothing scheme as well as the basic GETMe simultaneous scheme. It is emphasized once more that for these basic approaches, shape preservation cannot be expected during smoothing. It should also be noticed that both initial meshes are already of high initial quality, as can be seen by the mesh quality numbers given in Table 6.2.

(a) Flange mesh   (b) Stub axle mesh

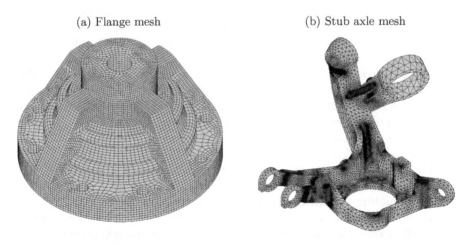

Figure 6.10: Initial meshes for unrestricted surface mesh smoothing.

Table 6.2: Initial mesh quality numbers and results of unrestricted surface mesh smoothing.

Mesh	Quadrilateral flange		Triangular stub axle	
	q_{min}	q_{mean}	q_{min}	q_{mean}
Initial	0.4489	0.9635	0.0055	0.9512
Basic Laplace	0.0333	0.8238	0.0001	0.4594
Basic GETMe sim	0.8638	0.9947	0.9003	0.9926

The quadrilateral flange model surface mesh consists of 15,154 nodes and 15,170 quadrilaterals, with a minimal element quality of $q_{min} = 0.4489$ and a mean mesh quality of $q_{mean} = 0.9635$. Its initial bounding box dimensions are given by 16, 16, and 9, respectively. Since the mesh is of already high quality, smoothing is expected to converge fast without extensive mesh modifications. Nevertheless, basic Laplacian smoothing performs 1,125 iterations, whereas basic GETMe simultaneous smoothing converges after only 233 iterations. In both cases, smoothing was terminated after the maximal node movement obtained within one smoothing step dropped below 1% of the mean edge length of all initial mesh edges. The resulting meshes of both smoothing approaches are depicted in Figure 6.11.

By comparing these meshes with the initial mesh, it can be seen in Figure 6.11a that basic Laplacian smoothing changes the shape and size of the mesh considerably. Mesh quality is significantly decreased, since quality aspects are not particularly addressed by this smoothing approach. Furthermore, due to neighboring node averaging, basic Laplacian smoothing has a strong defeaturing effect. That is, fundamental geometric characteristics are smoothed out and the resulting shape is significantly simplified.

In contrast, basic GETMe simultaneous smoothing almost preserves the initial shape, as can be seen in Figure 6.11b. Since mesh quality cannot be further improved within the surface of the initial mesh, elements are improved by moving nodes in the local surroundings, resulting in stick-out elements if compared to the original surface. Thus, basic GETMe smoothing has the characteristics of a locally acting, quality-improving wrinkling process of the surface mesh. Furthermore, being based on regularizing element transformations, quality improvement is an intrinsic property of basic GETMe smoothing. Therefore, node movement converges on a natural basis if perfect quality elements are achieved. This results in local modifications without altering the initial shape and dimension too much.

These characteristics are also reflected by the quality graph depicted in Figure 6.11c. Here, for each iteration of basic Laplacian smoothing and basic GETMe simultaneous smoothing, the resulting minimal element quality q_{min} and mean mesh quality q_{mean} are depicted with respect to the iteration number. In contrast to basic Laplacian smoothing, basic GETMe smoothing significantly increases both quality numbers within the first 75 iterations. After that, quality

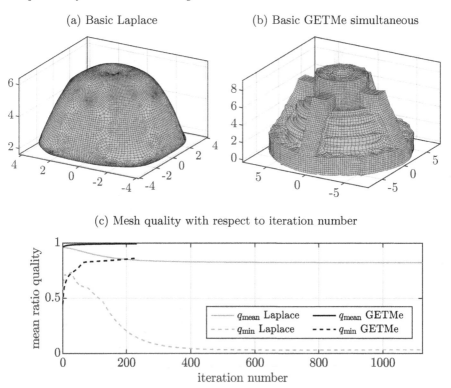

(a) Basic Laplace   (b) Basic GETMe simultaneous

(c) Mesh quality with respect to iteration number

Figure 6.11: Unrestricted smoothing of the quadrilateral flange model surface mesh.

numbers are only slightly improved. In contrast, both quality numbers are significantly decreased by basic Laplacian smoothing during the first 400 iterations. This holds, in particular, for the minimal element quality number.

Results for the stub axle surface mesh with bounding box dimensions 292.64, 280.36, and 145.45, respectively, are given in Figure 6.12. Here, the initial triangular mesh of the stub axle model depicted on the right of Figure 6.11 was likewise smoothed by basic Laplacian smoothing and basic GETMe simultaneous smoothing. In the case of the latter, the fixed parameters $\lambda = 0.35$ and $\theta = \pi/12$ have been used for element transformation. The stub axle mesh consists of 21,509 nodes and 43,086 triangular elements, and its initial quality numbers are given by $q_{\min} = 0.0055$ and $q_{\mean} = 0.9512$.

Basic Laplacian smoothing terminated after 1,385 iterations, while basic GETMe simultaneous smoothing terminated after only 331 iterations. As can be seen in Figure 6.12a, basic Laplacian smoothing again led to a significant change of shape and a reduction of size. According to Table 6.2, both quality numbers decreased considerably. In contrast, basic GETMe simultaneous smoothing again only led to local mesh modifications improving both quality

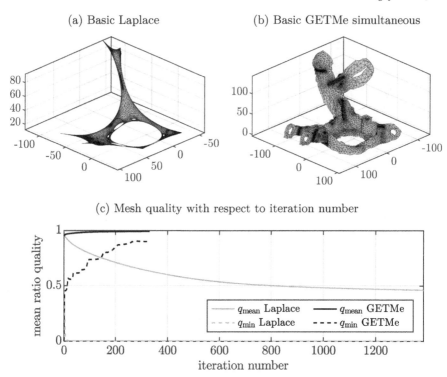

Figure 6.12: Unrestricted smoothing of the triangular stub axle surface mesh.

numbers. This is also reflected by the resulting geometric shape depicted in Figure 6.12b, which resembles that of the initial mesh. Thus, in the given examples the characteristics of basic GETMe smoothing are more favorable, both with respect to shape preservation and quality improvement if compared to basic Laplacian smoothing. Again, this is also visible in the quality with respect to iteration number graph depicted in Figure 6.12c.

The incorporation of back projection techniques into the GETMe smoothing process to preserve the shape of the initial model will be discussed in Section 6.4.1.

6.4   Extending the GETMe mesh improvement framework

In this section, outlines are given for extending GETMe smoothing algorithms by standard techniques such as local back projection, to preserve shape during surface mesh smoothing or topology modification approaches in order to overcome mesh topology-induced smoothing impediments. The latter can also be used in combination with basic GETMe simultaneous smoothing to untangle meshes. Furthermore, anisotropic mesh smoothing by using customized transformations and scaling schemes is addressed.

6.4.1 Surface mesh smoothing

So far, boundary nodes have been kept fixed during the smoothing process to preserve the shape of the underlying geometric model. However, mesh quality of volumetric meshes can further be improved if boundary nodes are also allowed to move. Here, movement must be restricted to the geometric model to preserve its geometric features. This is also relevant in the context of polygonal surface mesh smoothing. As has been demonstrated by Figure 5.15, iteratively applying the regularizing transformation to arbitrary polygons in three-dimensional space with non-coplanar nodes results in regular polygons with coplanar nodes. Thus, transforming single elements does not guarantee that the resulting nodes are still located on the initial geometric model. This also holds for the node averaging process of the GETMe simultaneous approach. Although GETMe has the tendency to preserve the initial model, as illustrated by Figure 6.11 and Figure 6.12, shape preservation is not guaranteed by the smoothing process. This also holds for any other smoothing technique. However, this can easily be overcome by using standard shape preservation techniques as applied in other smoothing methods.

A common approach is to apply the smoothing algorithm in its standard form and to conduct a projection technique afterwards, either after each iteration or after the entire smoothing process. In this projection step, each node is back projected on a prescribed reference model. This model can, for example, be based on an implicit analytic representation, as generated by a constructive solid geometry model approach, a piecewise parameterized surface model designed by a CAD approach, or an initial surface mesh without further geometric information.

For the first two cases, the mathematical model of the surfaces provides additional analytic information, such as surface tangents, normals, and differentiability information, which can be used for *back projection* by means of mathematical minimization. Since smoothing leads to only moderate node movements, the projected position of the previous smoothing iteration is already a good starting point for a local distance minimization of the new smoothed mesh node and its projection point on the reference surface. If this reference model is only given by the initial mesh, this results in a local search starting from the initial mesh element, which minimizes the distance to the smoothed node. In addition, face normals can also be derived from this initial mesh. They are used to judge whether the orientation and validity of the surface elements are maintained during smoothing.

To preserve the characteristics of the geometric model during smoothing and applying the back-projection technique, distinct types of projections are considered. These are related to the location of the mesh node with respect to *geometric features* like edges or corners of the model. If the reference model is a constructive model, features such as edges, which are defined by surface intersections, are usually classified during the construction process. In contrast, if only an initial mesh is given, such features must be derived by analyzing this mesh.

In the case of triangular surface meshes, a common technique is to derive this feature information by analyzing angles enclosed by neighboring triangular face normals. If an angle exceeds a given threshold, the common edge is denoted as *feature edge* and its two mesh nodes as *feature nodes*. A chain of consecutive feature edges is denoted as a *feature curve*. Finally, mesh nodes that belong to more than one feature curve are denoted as *feature corners*. The latter also holds for the start and end nodes of non-closed feature curves not connected to any other feature curve. In the case of higher-order polygonal elements, the elements are triangulated, and feature edges are detected by analyzing neighboring triangles of different elements.

Results of this approach are depicted in Figure 6.13 for the quadrilateral surface mesh of a flange and the triangular surface mesh of a stub axle already considered in the previous section. Here, an angle threshold of $\pi/6$ has been used to classify features. The resulting feature edges are marked by thick black lines, while feature corners are marked by black discs. The feature curves lead to a natural subdivision of the mesh into individual feature surfaces.

To optimally preserve the geometric features during smoothing, the following back projection techniques are used depending on the feature type of a mesh node:

○ Feature corner node: preserve geometric position, i.e., no movement

○ Feature curve node: project on associated feature curve, i.e., curve restricted movement

○ Feature surface node: project on associated feature surface, i.e., surface restricted movement

Hence, by projecting the smoothed node on the appropriate geometric entity, features and thus shape characteristics are preserved. Furthermore, it is ensured that feature surface nodes do not traverse feature curves, and thus they always belong to the same feature surface. Similarly, feature curve nodes do not traverse feature corner nodes.

If the initial mesh is used as a reference model, lookup tables are used to map each node to an initial mesh element on which the node is located. Subsequently, elements are transformed using the current element normals for polygonal element transformation according to Definition 5.7. After applying one smoothing step, the surface back projection can be implemented by finding the minimal distance element by a local search strategy starting at the previous minimal distance element taken from the lookup table. Element orientation and thus validity can be assessed after projection by comparing the reference

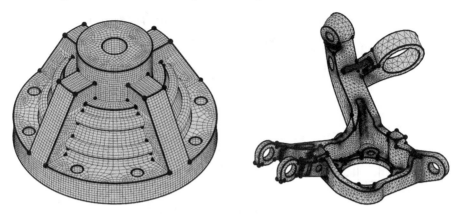

Figure 6.13: Quadrilateral and triangular surface meshes after feature detection.

normals with the resulting element normals. If invalid elements occur, the mesh is locally reverted to the configuration before this smoothing step. This is also the approach of the iterative node resetting technique of GETMe smoothing to guarantee mesh validity. Hence, for smoothing surface meshes by GETMe, node projection is simply applied before this resetting technique. After that, the lookup table is updated. This approach also applies to other smoothing methods.

6.4.2 Anisotropic mesh smoothing

Standard techniques for anisotropic mesh generation and smoothing, based, for example, on parametric mappings [93], can also naturally be applied to GETMe smoothing without further modifications. Alternatively, customized element transformations can be used by GETMe smoothing to control the smoothing objectives. This section describes concepts for smoothing anisotropic meshes based on this approach, which allows to control the target shape, size, and direction of mesh elements.

The first example considers basic GETMe smoothing using not only one transformation applied to all mesh elements, but different customized transformations depending on the mesh requirements. This is illustrated in the example of a simple mesh consisting of 13 quadrilateral elements of the same size, depicted in Figure 6.14a. The four nodes marked by circles are fixed during smoothing. The other 24 non-fixed nodes are marked by discs.

The number within each element represents a prescribed target edge length ratio to be achieved by the basic GETMe smoothing approach. Here, element widths increase starting from the boundaries towards the center of the mesh. Such specific mesh layer widths are required, for example, in specific computational fluid dynamics applications.

A customized rectangle transformation to achieve such a prescribed edge

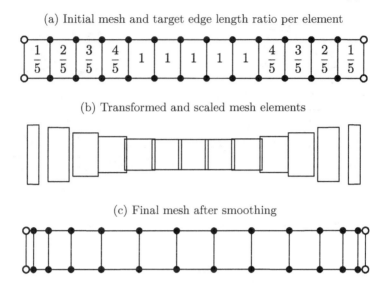

(a) Initial mesh and target edge length ratio per element

(b) Transformed and scaled mesh elements

(c) Final mesh after smoothing

Figure 6.14: Basic GETMe smoothing using element-specific transformations for shape control.

length ratio is given in Section 5.1.3. Using this together with the choice of eigenvalues $\eta_0 = 1$, $\eta_1 = 10$, and $\eta_2 = \eta_3 = 1/10$, each element is transformed by a transformation that depends on the target ratio. Applying edge length-preserving scaling after transforming each initial mesh element individually results in the intermediate elements depicted in Figure 6.14b. In accordance with (6.1), the new positions of non-fixed mesh nodes are computed as the arithmetic means of these intermediate nodes. Applying this scheme iteratively results in the mesh with desired element edge length ratios depicted in Figure 6.14c.

As in the case of regularizing transformations, the resulting mesh depends on the properties of the incorporated element transformations. However, in the case of more complex meshes, the new node computation approach based on the arithmetic mean of intermediate nodes has an additional influence and might result in averaging the shapes of neighboring elements. In a computational fluid dynamics context, the generation of boundary layers with prescribed width is usually more important than having almost regular elements in the core mesh. This can be achieved by using a weighted node averaging scheme for the new node position computation according to

$$p_i' := \frac{\sum_{j \in J(i)} w_{i,j} p_{i,j}'}{\sum_{j \in J(i)} w_{i,j}} \tag{6.7}$$

for non-fixed mesh nodes. In the case of meshes with boundary layers, it suffices to have a common weight for layer elements, which is larger than the common weight for the remaining kernel elements. This ensures that boundary layer

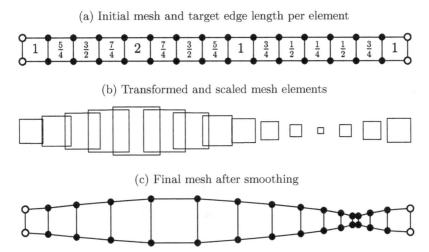

(a) Initial mesh and target edge length per element

(b) Transformed and scaled mesh elements

(c) Final mesh after smoothing

Figure 6.15: Basic GETMe smoothing using regularizing element transformations and element-specific scaling for size control.

elements preserve their desired shapes and are not drawn towards the core of the mesh. An example of this approach for a planar quadrilateral mesh with boundary layers is given in Section 7.1.3.1.

Meshes with almost regular elements of varying density are often used in an anisotropic mesh context. Here, smaller elements are used in regions where a strongly varying solution is expected, whereas larger elements are used in regions where only minor changes of the physical solution to be modeled are expected.

Various mesh generation methods exist for generating meshes with a pre-scribed element density [93]. For improving the resulting meshes, mesh smooth-ing methods ideally have two objectives. One is to improve mesh quality, the other to preserve or even generate the prescribed mesh density. The latter can be easily incorporated into the GETMe smoothing process by controlling the resulting transformed element size, as is depicted exemplarily in Figure 6.15.

The initial mesh shown in Figure 6.15a consists of 15 almost regular quadrilateral elements of width 67/60 and height one. Inside each element a target edge length is denoted, which should be achieved by smoothing. The sum of all these target edge lengths equals the mesh width 67/4. Fixed nodes of the mesh are marked by circles.

Since transforming an element usually alters its size, edge length-preserving scaling is applied within GETMe-based mesh smoothing. If a prescribed element size must be achieved, it is a natural step to scale the transformed element accordingly to achieve this prescribed size. The result of transforming each element of the initial mesh with a subsequent target edge length-based scaling step applied is depicted in Figure 6.15b. Here, the customized transformation with angle $\alpha = \pi/2$ and the same eigenvalues as in the previous example have

(a) Initial mesh and target direction angle per element

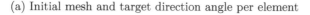

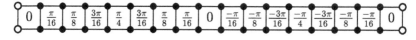

(b) Transformed, scaled, and rotated mesh elements

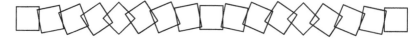

(c) Final mesh after smoothing

Figure 6.16: Basic GETMe smoothing using regularizing element transformations and element-specific direction control.

been applied. The resulting intermediate elements are used by basic GETMe simultaneous smoothing for new node determination based on intermediate node averaging.

Iteratively applying these steps results in the final smoothed mesh depicted in Figure 6.15c. As can be seen, the resulting element sizes almost match the prescribed values, where differences are due to different sizes prescribed for neighboring mesh elements. Hence, using target element size-based scaling allows to control mesh density both efficiently and effectively.

GETMe-based smoothing even allows to control the direction of the elements during mesh smoothing, as is illustrated in Figure 6.16. The initial mesh depicted in Figure 6.16a consists of 17 regular quadrilateral elements. As in the previous examples, the mesh is fixed along the left- and rightmost edge as is indicated by circular nodes.

Inside each element, a prescribed target angle between the x-axis and the lower element edge is given. Using the same customized transformation as in the previous target size example and rotating each element by the prescribed angle after scaling results in the intermediate elements depicted in Figure 6.16b.

Iteratively applying these transformation steps by basic GETMe simultaneous smoothing for intermediate element computation finally results in the mesh depicted in Figure 6.16c. The effect of following two smoothing objectives results in a bended mesh, which combines almost-regular elements and fulfilling of the prescribed direction angles.

As has been shown by these three examples, GETMe smoothing allows to easily incorporate target shape, size, and direction control for mesh elements. Whereas shape control can readily be achieved by using appropriate element transformations, size and direction control can both be incorporated into the subsequent transformed element scaling step. It is assumed that the target element size can be given as a scaling function $s : \mathbb{R}^2 \to \mathbb{R}$, and the target direction correction angle as a function $\delta : \mathbb{R}^2 \to \mathbb{R}$. A canonical point for evaluating these functions for a transformed element E'_j is its centroid $c'_j \in \mathbb{R}^2$. In doing so, the size and direction-corrected intermediate nodes $p_{i,j}^{(s)}$ of the element can be written as

$$p_{i,j}^{(s)} := s(c'_j)R(\delta(c'_j))(p'_{i,j} - c'_j) + c'_j,$$

where $R(\alpha)$ denotes the 2×2 rotation matrix for a given rotation angle α. Here, size control is achieved by scaling and direction control by rotation.

The angle correction function δ also usually depends on the actual direction of the element E'_j, which can be determined with respect to the target shape of the transformation as described in Section 5.1.3. An example of incorporating target-based element scaling into basic GETMe simultaneous for anisotropic mesh smoothing is given in Section 7.1.3.2.

6.4.3 Combining GETMe with topology modification

In this section, an approach for mesh improvement by combining smoothing with topology modification techniques is described in the example of planar triangular meshes. Other mesh types can be improved similarly, using the appropriate topological modification templates.

As has been demonstrated in Section 4.3.1, simple triangular mesh modification templates exist. These are namely the removal of nodes surrounded by three or four elements, node insertion techniques for the longest element edge, and edge flipping. These techniques are combined with simultaneous GETMe smoothing to iteratively improve element quality as well as mesh topology. Here, GETMe smoothing is not only used for mesh improvement but also as an indicator for topological deficiencies of the mesh. If GETMe converges, the mesh quality is either sufficient or topological problems should be resolved using the template techniques named before. An exemplary approach based on this scheme is given by Algorithm 8.

In line 2 the algorithm is preliminarily terminated if the lowest element quality is already larger than a given threshold. Otherwise, whether inner nodes surrounded by three or four elements exist is checked. If this is the case, the associated templates are applied. In doing so, nodes surrounded by three elements are removed from the mesh and the three triangles are replaced by the enclosing triangle. Nodes surrounded by four elements are also removed. However, the enclosing quadrilateral has to be split along one of its diagonals to result in two triangles replacing the four triangles sharing one common node. The selection of the proper diagonal can either be based on maximizing

---

Algorithm 8: Mesh improvement approach combining GETMe simultaneous smoothing and topological modifications.

Input   : Valid triangular mesh
Output : Improved valid triangular mesh

---

1 **for** *iter* := 1 **to** maxIteration **do**
2   **if** $q_{\min} >$ qualityThreshold **then**
3   **break**;
4   **end**
5   **if** *Inner nodes surrounded by 3 or 4 triangles exist* **then**
6   Apply node removal templates;
7   Apply GETMe simultaneous smoothing;
8   **end**
9   Determine lowest quality triangle T_ℓ and its longest edge e_ℓ;
10   **if** e_ℓ *is an outer edge of the mesh* **then**
11   Split triangle on edge midpoint;
12   **else**
13   Swap edge with e_ℓ-neighbor triangle;
14   **end**
15   Apply GETMe simultaneous smoothing;
16 **end**

---

the minimal element quality of the resulting two triangles, or on optimizing the resulting number of incident elements of the affected mesh nodes. The removal of nodes with four surrounding elements must be applied iteratively, if the nodes to be removed share common elements. After these modifications, simultaneous GETMe smoothing is applied.

In line 9 of the algorithm, the triangular element T_ℓ with the lowest mean ratio quality number is selected for improvement and its longest edge e_ℓ is determined. If the edge is part of the mesh boundary, i.e., it is only an edge of the triangle T_ℓ, its midpoint is used as a splitting point. By connecting the midpoint with the edge opposite node of T_ℓ, this element is split into two new elements replacing T_ℓ. If e_ℓ is shared by two elements, the edge is swapped. That is, the two triangles sharing e_ℓ are replaced by the enclosing quadrilateral split along the diagonal, which is not e_ℓ. As in the case of the node removal modifications, GETMe simultaneous smoothing is applied subsequently. These steps are iteratively applied until a prescribed maximal iteration number is reached.

It should be noticed that GETMe simultaneous is used for smoothing instead of the combined GETMe approach. This is because iteratively applying GETMe smoothing results in GETMe simultaneous to be applied after GETMe sequential. Since GETMe simultaneous is focused on improving the overall mesh quality, it would weaken the local improvements obtained by the preceding GETMe sequential step. Furthermore, the element of lowest quality is topologically modified to improve mesh quality after smoothing. Since GETMe simultaneous already leads to satisfactory results with respect to q_{min}, this combined approach of smoothing and topological modifications results in high-quality meshes. Nevertheless, GETMe sequential can be applied after conducting Algorithm 8 to further improve the minimal element quality.

This combined mesh improvement algorithm is applied to the initial mesh depicted in Figure 6.17a. The mesh has been generated by Delaunay triangulation of 100 random nodes placed in a unit square, resulting in 185 triangles. The number of incident elements per node range from two to ten and the initial mesh quality numbers are given by $q_{min} = 0.0056$ and $q_{mean} = 0.6405$. In addition, the mesh contains inner nodes with three or four surrounding elements. Examples of such configurations are marked by thick black lines in Figure 6.17a. The minimal element quality threshold for preliminary termination of the algorithm was set to the mean ratio value 0.9. To demonstrate the full potential of the combined mesh improvement approach, boundary nodes have not been kept fixed during smoothing.

Applying the node removal templates and GETMe simultaneous smoothing results in the mesh depicted in Figure 6.17b. Due to the topological modifications, the number of nodes and elements decreased to 75 and 135, respectively. In contrast, due to these modifications and applying mesh smoothing, the mesh quality numbers already increase to $q_{min} = 0.6668$ and $q_{mean} = 0.9064$. According to the combined algorithm, the next step consists of picking the element with the lowest quality number, which is marked by thick black lines in Figure 6.17b. As can be seen, the longest edge of this element is a boundary edge of the mesh. Hence, the triangle is simply split into two triangles along the black dashed line. After that, smoothing is applied again, completing the first iteration of the algorithm.

Figure 6.17c depicts the results after applying 11 iterations. Here, the element with the lowest quality number $q_{min} = 0.8841$ marked by thick black lines shares its longest edge with another element. Hence, edge swapping is applied, resulting in the two triangles separated by the black dashed line. As can be seen, these elements are of lower quality. However, for two neighboring nodes, the number of incident elements changes to the ideal value six, which allows the subsequent smoothing step to achieve better results.

The result depicted in Figure 6.17d is obtained after 15 iterations of the algorithm. This mesh consists of 77 nodes and 132 triangles with a minimal element quality of $q_{min} = 0.9215$ and a mean mesh quality of $q_{mean} = 0.9637$. The maximal number of incident elements per inner node has been decreased to seven, which improves smoothing results significantly. Applying GETMe simultaneous without incorporating topological modifications to the initial mesh would result in the quality numbers $q_{min} = 0.4059$ and $q_{mean} = 0.8321$. Thus, modifying mesh topology effectively resolves the impediments for mesh smoothing.

The given algorithm can be improved in several ways. For example, instead of only selecting the lowest quality element for topological modifications, all

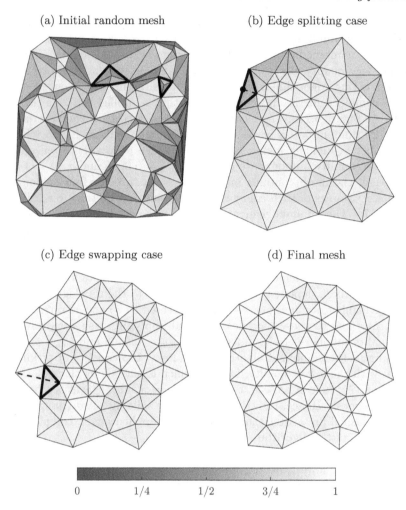

Figure 6.17: Combined mesh improvement approach applied to a random nodes Delaunay mesh.

elements below a given mean ratio quality threshold could be selected. For each element, the appropriate template is applied, if it does not interfere with the modification template of other selected elements. This increases the rate of topological improvements with respect to smoothing iterations. In addition, iterative longest-edge refinement techniques controlling element size could also be applied, resolving the problem of a deadlock caused by two neighboring elements with alternately swapping diagonals. Furthermore, the given algorithm can be combined with local point insertion and deletion techniques based on the Delaunay criterion. However, it should be noticed that smoothing does not guarantee the preservation of this criterion.

6.4.4 Using GETMe for mesh untangling

The combined mesh improvement approach incorporating GETMe simultaneous smoothing as well as topological modifications requires the initial mesh to be valid. The same holds for the simultaneous, sequential, and adaptive GETMe smoothing approach. This is due to the incorporation of element and mesh quality numbers for transformation and termination control. Hence, invalid meshes do not fulfill the prerequisites of these algorithms to be applicable.

In contrast, mesh validity is not a prerequisite of the basic GETMe simultaneous approach, since it is only based on element transformations using fixed parameters and a geometric termination criterion. Furthermore, as has been shown by the examples depicted in Figure 6.2 and Figure 6.5, applying basic GETMe simultaneous smoothing has a strong untangling effect. However, since smoothing cannot overcome topological deficiencies and due to its pure geometric approach, validity of the resulting mesh is not guaranteed.

Therefore, it is a natural approach to apply basic GETMe simultaneous smoothing first, to try to *untangle* a given mesh, i.e., to turn its invalid parts into valid parts. If the resulting mesh is still invalid, topological modifications must be applied as a second step. The easiest way of doing so is to remove the remaining invalid parts of the mesh and to fill the resulting gaps by remeshing them. This approach is summarized by Algorithm 9.

Algorithm 9: GETMe-based mesh untangling.

Input   : Invalid mesh
Output: Valid mesh

1 Apply basic GETMe simultaneous smoothing;
2 **if** *mesh is not valid* **then**
3   Remove invalid submeshes;
4   Mesh regions of removed invalid submeshes;
5 **end**

Due to construction, the resulting mesh is valid. Furthermore, due to applying basic GETMe simultaneous smoothing in the first stage, the not-remeshed parts are expected to be of good quality. Nevertheless, if the mesh obtained after remeshing the invalid parts does not fulfill the quality requirements, all other GETMe smoothing and improvement approaches based on valid meshes as described in this chapter can now be applied, to result in a high quality mesh.

7

Numerical examples

This chapter provides a broad variety and an in-depth analysis of numerical results obtained by smoothing meshes of various types. Results are given for GETMe smoothing variants, some variants of Laplacian smoothing, and a state-of-the-art global optimization-based method. Synthetical as well as real-world models and meshes are considered. Some examples are based on those presented in the publications [143, 153, 156, 159, 160, 162, 163].

Results have been obtained by a straightforward C++ [132] implementation of GETMe and Laplacian smoothing variants. They are compared to the results of the feasible Newton-based global optimization approach of the Mesh Quality Improvement Toolkit (Mesquite) [99] version 2.3, which is also implemented in C++. Here, the state-of-the-art global optimization algorithms of Mesquite are based on optimizing the mean ratio quality number, as described in Section 4.2.2.

For all algorithms, runtimes have been measured on a personal computer with an Intel® Core™ i7-3770 CPU (quad core, 8 MB level 3 cache, 3.4 GHz frequency), 12 GB RAM, 64-bit Linux operating system with kernel 4.4.36, and the GNU C++ compiler version 6.1.1. HyperThreading™ and SpeedStep™ technology have been disabled.

7.1 Polygonal meshes

7.1.1 Mixed planar mesh

The first example considers a planar mixed mesh of a Mercator projection-based map mainly focusing on Europe. Here, smaller islands have been omitted, resulting in 168 unconnected domains as depicted in Figure 7.1. The quad dominant planar mesh of all domains consists of 92,480 triangles and 452,182 quadrilaterals, hence 544,662 elements in total, sharing 508,280 nodes. All 19,380 boundary nodes have been kept fixed during smoothing. Since the entire mesh is too fine to be fully depicted here, zoomed regions of Iceland and Norway are shown later.

To increase the smoothing potential, the initial mesh has been distorted by validity-preserving node movements. That is, inner mesh nodes have been

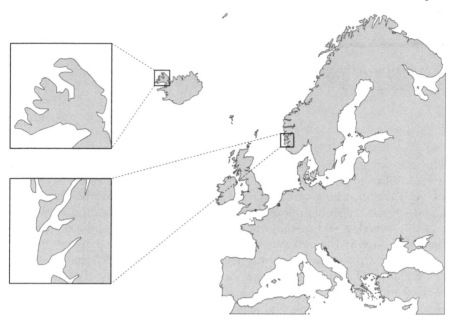

Figure 7.1: Domains of the Europe mesh with zoomed regions of Iceland and Norway.

iteratively moved by random translations. If such a translation led to invalid incident elements with respect to the mean ratio quality criterion, the translation has been reverted. This distortion process has been repeated, until the mean ratio quality number q_{mean} of the initial mesh dropped below 0.5. The lowest element quality number q_{min} of the mesh is given by 0.0006. Hence, the initial mesh is of very low quality with respect to both quality numbers, but still consists of valid elements only.

For comparison reasons, this initial mesh has been smoothed by smart Laplacian smoothing as described in Section 4.2.1. This, in contrast to standard Laplacian smoothing, preserves mesh validity. As a further benchmark, the global optimization-based approach provided by Mesquite has also been applied to the initial mesh. Both results are compared with those of the combined GETMe smoothing approach as described in Section 6.2.1. For the latter, the regularizing polygon transformation according to Definition 5.1 has been used. Here, transformation parameters have been set to the fixed values $\lambda = 1/2$ for all element types, $\theta = \pi/4$ for triangles, and $\theta = \pi/6$ for quadrilaterals in both the simultaneous and the sequential substep of GETMe smoothing. Mean edge length-preserving scaling has been applied and, for reasons of simplicity, the relaxation parameter of GETMe simultaneous has been set to $\varrho = 1$ and the weight exponent to $\eta = 0$. In the case of GETMe sequential smoothing, the penalty correction values for invalid elements, repeatedly selected elements and successfully transformed elements have been set to $\Delta \pi_i = 10^{-4}$, $\Delta \pi_r = 10^{-5}$, and $\Delta \pi_s = 10^{-3}$, respectively.

GETMe and smart Laplacian smoothing were terminated if the mean mesh quality number q_{mean} of two consecutive iteration meshes dropped below 10^{-4}. In the case of the GETMe sequential substep, mesh quality was evaluated after each cycle consisting of 100 single element transformations, and smoothing was terminated after 20 cycles without q^*_{min} improvement. Results of all smoothing approaches are given in Table 7.1, providing iteration, runtime, and quality numbers.

Table 7.1: Europe mesh smoothing results.

Method	Iter	Time (s)	q_{min}	q^*_{min}	q_{mean}
Initial	–	–	0.0006	0.0006	0.4789
Smart Laplace	25	37.82	0.0037	0.0037	0.9490
Global optimization	84	81.38	0.0021	0.0021	0.6753
GETMe	48 / 2,200	42.84	0.3332	0.3342	0.9610

As can be seen, smart Laplacian smoothing terminated after 25 iterations taking 37.82 seconds in total. However, during smoothing, 148,113 node updates had to be reverted, since otherwise invalid elements would have been generated. This large number of node resets also leads to an increased runtime per iteration, since a large number of element quality numbers had to be reevaluated. Nevertheless, the resulting mesh is of high quality with respect to the mean mesh quality number, resulting in 0.9490 but not with respect to the lowest element quality number, 0.0037.

Due to the complexity and severe distortion of the initial mesh, the global optimization approach was not able to result in better mesh quality numbers after 84 iterations taking 81.38 seconds in total. Using its own gradient norm-based termination criterion with a tolerance set to 10^{-6}, the mean mesh quality difference between the last but one and final iteration dropped to 0.0017.

In contrast, GETMe smoothing resulted in much better quality numbers with respect to both mean and minimal element quality, as can be seen in Table 7.1. In this, iteration numbers are given for the simultaneous as well as the sequential substep. The latter improved the minimal element quality 0.0133 after simultaneous smoothing to 0.3342 within the first 200 single element transformations, which shows the effectivity of the combined approach. Furthermore, GETMe is the only smoothing method that improved the minimal improvable element quality number $q^*_{min} = 0.3342$ beyond the non-improvable minimal fixed element quality number given by $q_{min} = 0.3332$. In contrast to smart Laplacian smoothing, the regularizing element transformation approach led in total only to 396 invalid elements, which had to be reset during smoothing. Hence, although the number of GETMe iterations in this example is larger than the number of smart Laplacian smoothing iterations, the total GETMe smoothing time of 42.84 seconds is only moderately increased.

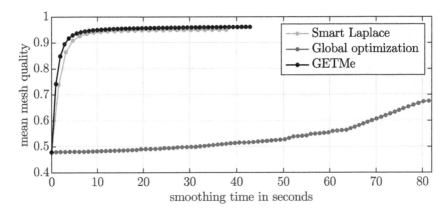

Figure 7.2: Europe mean mesh quality q_{mean} with respect to smoothing time.

Figure 7.2 depicts the mean mesh quality number q_{mean} after each mesh iteration with respect to the accumulated smoothing time. As can be seen, smart Laplacian smoothing and GETMe smoothing already lead to a significant improvement of mesh quality within the first few smoothing steps. In contrast, global optimization improves q_{mean} only slightly within the first iterations. As in the case of other optimization problems, the result of the global optimization-based approach depends on the quality of the start configuration, that is, the quality of the initial mesh.

Although smart Laplacian smoothing leads to comparable results with respect to the mean mesh quality if compared to GETMe smoothing, the resulting mesh is less suitable for finite element computations. This is because the resulting mesh obtained by smart Laplacian smoothing still contains 394 elements with quality numbers below 0.3, which may affect finite element solution efficiency and accuracy.

This is also obvious from the detailed mesh zooms depicted in Figures 7.3 and 7.4. These show zoomed regions of parts of Norway and Iceland, respectively, which are marked by black rectangles in Figure 7.1. In these zoomed figures, all mesh elements are shaded according to their mean ratio quality number, ranging from zero (invalid, dark gray) to one (ideal, light gray).

As can be seen on the upper left of each figure, the distorted initial mesh contains a large number of mid- and low-quality elements. Below each initial mesh, the result of global optimization is depicted, which is of improved but not satisfying quality. In contrast, the meshes obtained by smart Laplacian smoothing and GETMe smoothing depicted on the right consist of a large number of high quality elements. However, clusters of low quality elements are visible in all smart Laplacian smoothing meshes depicted on the upper right. GETMe smoothed meshes depicted on the lower right show the best result with respect to both quality numbers and are hence well suited for finite element computations.

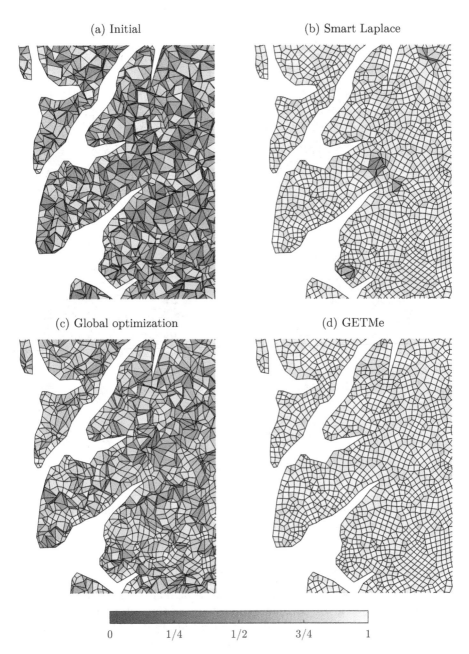

Figure 7.3: Europe mesh zoom to Norway. Mesh elements are shaded according to their mean ratio quality number.

(a) Initial   (b) Smart Laplace

(c) Global optimization   (d) GETMe

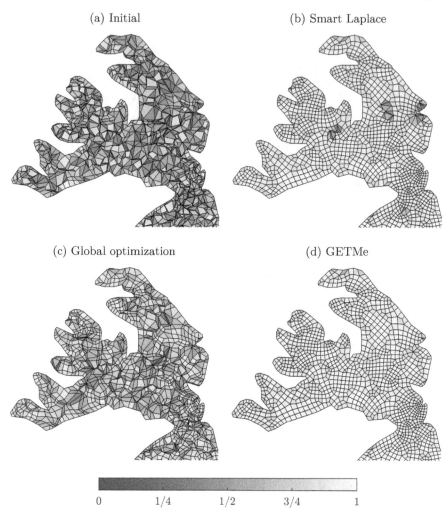

Figure 7.4: Europe mesh zoom to Iceland. Mesh elements are shaded according
to their mean ratio quality number.

7.1.2   Single element type surface meshes

The smoothing of polygonal surface meshes is demonstrated by the example
of two meshes. One is a triangular surface mesh of the piston model depicted
in Figure 7.5a, which is provided as a STEP-file by the Drexel University
Geometric & Intelligent Computing Laboratory model repository [35]. The
other is a quadrilateral surface mesh of the pump carter model depicted in
Figure 7.5b, of which a STEP-file was provided courtesy of Rosalinda Ferrandes,
Grenoble INP, by the AIM@SHAPE Mesh Repository [1].

   In both cases, comparably coarse meshes have been used for two reasons.

(a) Piston (b) Pump carter

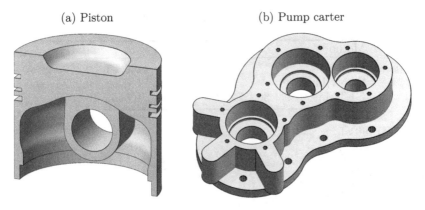

Figure 7.5: Models for surface meshes.

One reason is to make smoothing more challenging by a larger ratio of feature nodes, the other reason is the possibility of visualizing the entire resulting meshes.

Shape preservation has been ensured by applying the back-projection technique as described in Section 6.4.1. That is, after applying one smoothing step, nodes are back-projected on a reference background mesh of the model. Here, fixed feature nodes are preserved, feature edge nodes are projected on the associated feature curve consisting of a chain of feature edges, and all other nodes are projected on their associated feature surface. The same projection technique has been applied in the case of smart Laplacian and GETMe smoothing. Since the global optimization approach of Mesquite does not provide the same projection technique out of the box, results are not available for comparison.

Results for the piston model are considered first. The triangular surface mesh consists of 7,687 nodes and 15,374 elements. Since the surface is closed, there are no boundary nodes. However, 282 of the nodes are fixed feature nodes and 1,530 nodes belong to feature curves. The remaining 5,875 nodes are allowed to move within feature surfaces bounded by the feature curves connecting feature nodes and fixed corner nodes. Here, feature edges and nodes have been detected by analyzing the angle between elements sharing an edge. If such an angle exceeds $\pi/6$, the associated edge and nodes are classified as feature edges. If the angle between two feature edges exceeds $\pi/4$, the common node is classified as a fixed feature node.

As in the planar case, the initial mesh has been additionally distorted. The resulting mesh quality numbers are given in Table 7.2. Again, the initial mesh is of low quality with respect to both quality numbers q_{min} and q_{mean}. This is reflected by Figure 7.6a showing the initial mesh with elements shaded according to their mean ratio quality number. This mesh has been smoothed by smart Laplacian smoothing as well as GETMe smoothing using the same transformation and termination parameters as described in the previous section.

Table 7.2: Piston surface mesh smoothing results.

Method	Iter	Time (s)	q_{min}	q_{mean}
Initial	–	–	0.0001	0.3999
Smart Laplace	18	0.85	0.0008	0.9593
GETMe	36 / 4,200	1.04	0.6435	0.9600

(a) Initial (b) Smart Laplace (c) GETMe

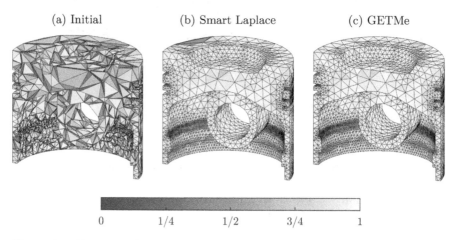

0   1/4   1/2   3/4   1

Figure 7.6: Piston surface mesh. Mesh elements are shaded according to their mean ratio quality number.

Compared to the case of smoothing the planar Europe mesh, only the additional back-projection step must be conducted for surface mesh smoothing to preserve the shape of the model.

Smart Laplacian smoothing was terminated after 18 iteration steps taking 0.85 seconds in total. The mean mesh quality was significantly improved from the initial value 0.3999 to 0.9593. In contrast, the resulting minimal element quality of 0.0008 is insufficient for subsequent finite element computations. As in the case of the planar Europe mesh, clusters of low quality elements remain as is visible in Figure 7.6b. During smoothing, 2,102 node updates had to be discarded to preserve mesh validity. Here, the orientation and thus the validity of the elements after one smoothing step was determined by comparing the resulting element normal with that of the reference mesh used for back-projection.

The number of GETMe simultaneous iterations amounting to 36 is significantly larger if compared with smart Laplacian smoothing. However, due to the reduced number of 465 element resets, smoothing is faster per iteration step due to the reduced number of element quality evaluations, which are more expensive due to the additional comparison of the element normals. The subsequent GETMe sequential substep improved the minimal element quality number to 0.6435 within 4,200 single element transformation steps. That is, in

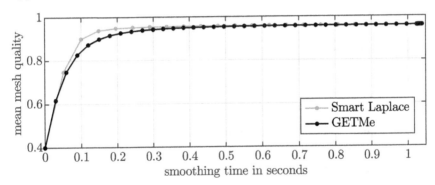

Figure 7.7: Piston mean mesh quality q_{mean} with respect to smoothing time.

contrast to smart Laplacian smoothing, GETMe smoothing removes effectively all low quality element clusters as can be seen in the resulting mesh depicted in Figure 7.6c. Furthermore, the simple back-projection technique used for shape preservation is well suited for both smoothing methods.

For the given example, both smoothing methods lead to comparable mean mesh quality numbers and have a similar runtime behavior with respect to q_{mean}. This is shown in Figure 7.7 depicting the mean mesh quality number with respect to smoothing time. Here, each marker represents the results obtained by one smoothing step.

The quadrilateral surface mesh of the pump carter model depicted in Figure 7.5b consists of 9,820 nodes and 9,836 elements. Feature nodes have been determined as in the case of the triangular surface mesh. To do so, the quadrilaterals have been converted into triangles by diagonal splitting. Such a triangular mesh has also been used for back-projection to preserve the shape of the model. The closed quadrilateral mesh consists of 929 fixed feature nodes, 1,157 feature curve nodes, and 7,734 feature surface nodes.

The quadrilateral mesh has been smoothed by smart Laplacian smoothing and GETMe smoothing using the polygonal transformation according to Definition 5.7. Here, the reference normal required for the transformation has been derived from the triangular reference mesh also used in the back-projection step. Smoothing results are given in Table 7.3. As in the case of the piston model, smart Laplacian smoothing fails to improve the minimal element quality considerably. Here, 147 elements of the smart Laplacian smoothing mesh have an element quality number below the minimal element quality number 0.6016 obtained by GETMe smoothing. Again, this is visible by the various clusters of low quality elements depicted in Figure 7.8b.

In contrast to the low minimal element quality number, the resulting mean mesh quality number 0.9666 of smart Laplacian smoothing is almost ideal, as in the case of GETMe smoothing, resulting in $q_{\text{mean}} = 0.9763$. The high quality results of GETMe smoothing with respect to both quality numbers is also visible in the element quality shaded mesh depicted in Figure 7.8c. During

Table 7.3: Pump carter surface mesh smoothing results.

Method	Iter	Time (s)	q_{min}	q_{mean}
Initial	–	–	0.0001	0.4000
Smart Laplace	31	1.96	0.0034	0.9666
GETMe	49 / 3,800	1.58	0.6016	0.9763

(a) Initial   (b) Smart Laplace   (c) GETMe

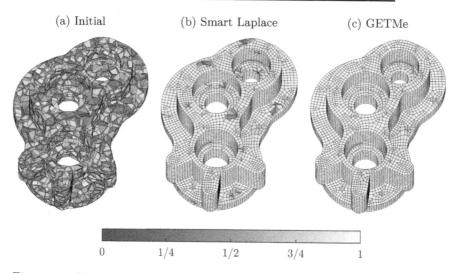

0   1/4   1/2   3/4   1

Figure 7.8: Pump carter surface mesh. Mesh elements are shaded according to their mean ratio quality number.

smart Laplacian smoothing, 6,654 node updates had to be discarded. In the case of GETMe smoothing, 2,002 elements had to be reverted to preserve mesh validity. The resulting graphs of mean mesh quality numbers with respect to smoothing time are depicted in Figure 7.9.

In the given and various additional smoothing experiments, it has been observed that smart Laplacian smoothing is well suited to improving the mean mesh quality of polygonal meshes. However, depending on the initial mesh quality, smart Laplacian smoothing often fails to significantly improve the minimal element quality. Both quality aspects are addressed by the GETMe smoothing approach by combining the simultaneous smoothing substep, focusing on improving mean mesh quality, with the sequential substep, particularly improving minimal element quality.

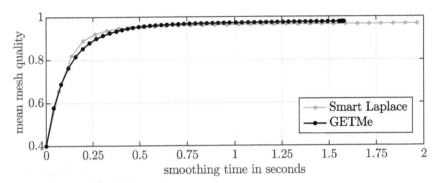

Figure 7.9: Pump carter mean mesh quality q_{mean} with respect to smoothing time.

7.1.3 Anisotropic meshes

7.1.3.1 Boundary layers preservation

In this section an example of smoothing a quadrilateral mesh of a star-shaped domain with a few boundary element layers of prescribed thickness will be considered. Such requirements occur, for example, in computational fluid dynamics applications using a specific turbulence model near the boundary of the simulation domain. The smoothing approach used follows the concept described in Section 6.4.2.

The boundary of the domain depicted in Figure 7.10a is parameterized by the complex-valued curve

$$z(\beta) := \left(1 + \frac{3}{10}\cos(5\beta)\right) \mathrm{e}^{\mathrm{i}(\beta+\pi/10)}, \text{ where } \beta \in [0, 2\pi).$$

By choosing 150 equidistant values for the angle β, boundary nodes and the associated polygonal approximation of the domain have been derived. It should be noticed that due to the parameterization, the density of the boundary nodes varies along the boundary. Starting from the polygonal approximation, five layers of quadrilateral elements with increasing thickness have been constructed inwards. Here, the element edge length ratios amount to $k/5$ with $k \in \{1, \ldots, 5\}$. The remaining part of the domain has been meshed by a paving approach. A part of the resulting mesh is depicted in Figure 7.10b.

After this, all interior node positions of the resulting mesh have been replaced by random values $a + b\mathrm{i}$ with $a, b \in [-1.3, 1.3]$ to demonstrate the ability of basic GETMe simultaneous smoothing to recover the prescribed boundary layers and to result in high-quality core meshes if there is a suitable mesh topology. This results in the severely distorted initial mesh depicted in Figure 7.10c, consisting of 1,228 nodes and 1,152 quadrilateral elements, of which 992 are invalid due to the random inner node positions.

(a) Domain and boundary nodes   (b) Undistorted mesh (part)

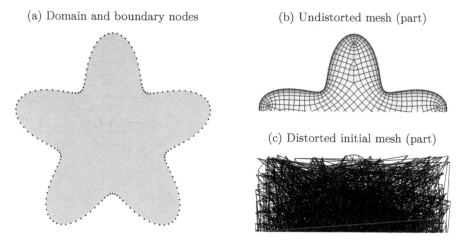

(c) Distorted initial mesh (part)

Figure 7.10: Star mesh domain with boundary nodes and meshes.

In Section 5.1.3 a customized transformation has been presented, which, if applied iteratively, results in quadrilateral elements with a prescribed edge length ratio. Here, the entries of the transformation matrix depend on the angle α enclosed by the diagonals of the target element. This transformation has been used with α chosen individually per mesh layer according to the values given in Table 7.4. In addition, the eigenvalues have been set to $\eta_0 = 1$, $\eta_1 = 10$, and $\eta_2 = \eta_3 = 1/10$ to ensure a fast convergence. Since the direction of the elements is prescribed by the fixed boundary nodes, a rotation correction according to (5.31) was not applied.

Table 7.4: Star mesh element transformation target parameters.

Layer	Edge length ratio	Angle α	Mean ratio
1	1/5	0.3948	0.3846
2	2/5	0.7610	0.6897
3	3/5	1.0808	0.8824
4	4/5	1.3495	0.9756
5 and inner	1	$\pi/2$	1

The edge length ratios given in the table match those of the mesh generation process. That is, smoothing using these layer-specific element transformations aims to recreate the boundary layers of increasing element thickness. Due to the ratio one, the fifth layer consists of almost regular elements. Therefore, the associated transformation matrix has also been used for transforming the remaining inner mesh elements. The last column also gives the mean ratio number of the ideal target element for each layer. To enforce the generation of the desired mesh layer thickness, the node weights of the transformed layer

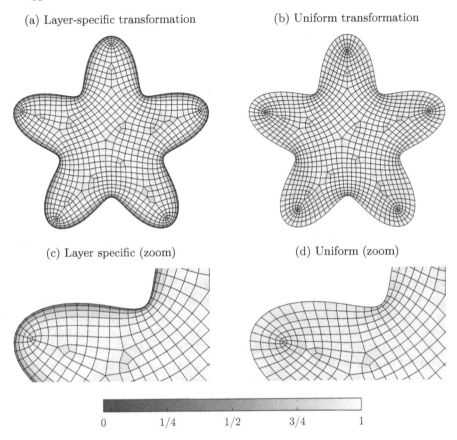

(a) Layer-specific transformation   (b) Uniform transformation

(c) Layer specific (zoom)   (d) Uniform (zoom)

0   1/4   1/2   3/4   1

Figure 7.11: Star mesh example smoothed by basic GETMe simultaneous.

elements have been set to ten and those of the inner mesh elements to one in the weighted node averaging computation scheme according to (6.7).

Figure 7.11a depicts the resulting mesh obtained by applying basic GETMe simultaneous smoothing, using the layer-specific element transformations, to the distorted initial mesh. The smoothing process was terminated after the maximal node movement dropped below 10^{-4}, which resulted in 388 iterations due to this restrictive threshold. For comparison reasons, the distorted initial mesh has also been smoothed by basic GETMe simultaneous smoothing using the transformation defined by $\alpha = \pi/2$ for all elements of the mesh and all node weights set to one. That is, by the standard approach used for regularizing all mesh elements. The results are depicted in Figure 7.11b.

The effect of using layer-specific element transformations are clearly visible in the elements shaded according to their mean ratio quality number. Whereas using layer-specific element transformations results in the desired layers of increasing edge length ratios, the uniform transformation approach leads to

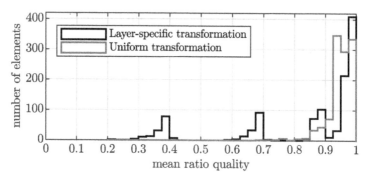

Figure 7.12: Star mesh histograms of element mean ratio quality numbers.

a uniform mesh quality with a few exceptions near the concave boundary parts. This is reflected by the zoomed meshes depicted in Figure 7.11c and Figure 7.11d. In the case of the layer-specific transformation, the resulting elements of one layer using the same transformation are of similar quality.

This is also visible from the element quality histograms depicted in Figure 7.12 for both cases. In the case of the layer-specific transformation, one can observe a cluster of elements near each mean ratio value given in the last column of Table 7.4, since applying basic GETMe simultaneous smoothing results in elements similar to the requested target shapes. In the case of the uniform transformation, the mesh quality numbers result in $q_{min} = 0.6864$ and $q_{mean} = 0.9512$ with a uniform distribution of high quality elements.

Due to the severely distorted initial mesh, it has been demonstrated that the resulting configuration after smoothing is not only a matter of the initial mesh and its topology. Instead, the characteristics of the resulting mesh can easily be controlled by choosing transformations resulting in the desired target shapes.

7.1.3.2 Mesh density preservation

In this section, an anisotropic triangular NASA mesh of an airfoil with two extended flaps is considered. The original mesh depicted in Figure 7.13a is contained as sample data in the numerical computing environment MATLAB® by MathWorks® Inc. [97]. The mesh consists of 8,034 triangular elements sharing 4,253 nodes. The 425 nodes defining the shape of the airfoil and the two flaps have been kept fixed during smoothing to preserve the shape of the model. In Figure 7.13a, the full mesh is depicted on the left and a zoomed version on the right side. In both cases, mesh elements are shaded according to their mean ratio quality number.

The original mesh has been scaled to fit into a unit square. The resulting element perimeters range from 0.0006 to 0.3503, and differ by a factor of about 586. For each triangle E_j, the associated centroid c_j and perimeter s_j representing the sum of all element edge lengths has been derived. This

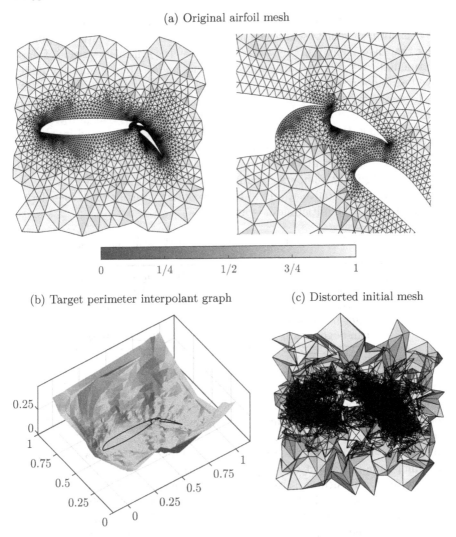

(a) Original airfoil mesh

0   1/4   1/2   3/4   1

(b) Target perimeter interpolant graph   (c) Distorted initial mesh

Figure 7.13: Anisotropic airfoil mesh example.

data has been used to compute a piecewise linear scattered data interpolant, which maps a centroid position to the associated target element perimeter. The resulting interpolant is depicted in Figure 7.13b with airfoil and flap contours marked by black lines.

This interpolant was used by basic GETMe simultaneous smoothing as a target scaling function to scale the transformed elements, so as to preserve the initial mesh density during smoothing. To be precise, in accordance with the scheme described in Section 6.4.2, each element was first transformed by a regularizing triangle transformation, its centroid was computed, the target perimeter was derived by evaluating the interpolant at the given centroid, and the transformed element was scaled in order achieve the given target perimeter. After that, new node positions were computed as the arithmetic means of these intermediate transformed and scaled element nodes. Here, boundary nodes have been kept fixed.

To demonstrate the potential of basic GETMe simultaneous smoothing and the preservation of the initial mesh density by applying target-based element scaling, the original mesh has additionally been distorted by random node movements, resulting in the mesh depicted in Figure 7.13c. The element perimeters of the distorted mesh vary from 0.0022 to 0.6164. As can be seen by the mesh quality numbers given in Table 7.5, the original mesh is already of high quality. In contrast, due to random node distortions not considering quality aspects, the initial mesh is invalid.

Table 7.5: Airfoil initial and smoothed mesh qualities.

Method	Iterations	q_{min}	q_{mean}
Original	–	0.2680	0.9603
Distorted initial	–	invalid	invalid
Target-based scaling	287	0.2910	0.9714
Mean edge length scaling	403	0.3027	0.9744

The initial mesh was smoothed by basic GETMe simultaneous smoothing using the perimeter interpolant for element scaling, as well as basic GETMe simultaneous smoothing using mean edge length-preserving scaling for comparison reasons. Results for both smoothing methods are also given in Table 7.5. As can be seen, the resulting mesh quality is improved in both cases if compared to the undistorted original mesh. Both algorithms were terminated after the maximal node movement dropped below the tight tolerance of 10^{-4}.

The mesh obtained by basic GETMe simultaneous smoothing using target-based element scaling is depicted in Figure 7.14a. By comparing the images with those of the original mesh depicted in Figure 7.13a, it can be seen that the smoothed mesh density matches that of the initial mesh. In addition, the number of lower-quality elements is reduced, as is also visible in the zoomed images shown on the right side of both figures.

As mentioned before, the initial distorted mesh has also been smoothed by the basic GETMe simultaneous smoothing scheme using the classic edge length-preserving scaling approach. That is, after transforming an element has been scaled such that the transformed element perimeter matches the element perimeter before transforming the element. The resulting mesh is depicted in Figure 7.14b. By comparing the results for both scaling approaches, it can be seen that the meshes only differ slightly. The reason for this is the high-quality original anisotropic mesh, as will be explained in the following.

Generating an anisotropic mesh with almost regular elements has an impact on mesh topology. Since basic GETMe simultaneous smoothing using

(a) Basic GETMe simultaneous smoothed mesh with target-based element scaling

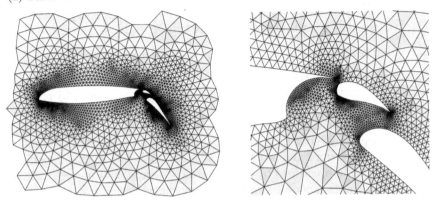

(b) Basic GETMe simultaneous smoothed mesh with edge length-preserving scaling

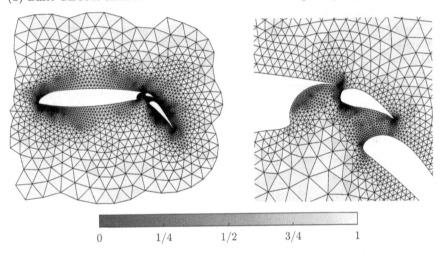

Figure 7.14: Smoothed airfoil mesh.

regularizing element transformations aims to generate regular mesh elements, the resulting mesh has a density similar to the original mesh. Furthermore, the fixed boundary nodes and the associated edges provide a fixed sizing information for the boundary elements, which propagates through the entire mesh. Hence, using regularizing element transformations results in an intrinsic support for anisotropic mesh smoothing without the need for specific element-scaling schemes in this case. Nevertheless, target diameter-based scaling is recommended for anisotropic mesh smoothing if a reasonable initial mesh quality or topology cannot be expected and a mesh-density target function is at hand.

7.2   Single element type volumetric meshes

7.2.1   Tetrahedral meshes

Tetrahedral meshes are the most common volumetric meshes, since Delaunay-based meshing methods are robust and efficient for meshing arbitrarily complex models. Hence, this section considers two tetrahedral meshes of real-world models. One is the modular hip endoprosthesis depicted on the left of Figure 7.15a. It was developed by NIKI Ltd. within the research project SKELET funded by the third European Framework program [139,154]. The isotropic modular endoprosthesis mesh consists of 13,192 elements and 2,669 nodes. 1,228 of these nodes are surface nodes, which will be kept fixed during smoothing. That is, the surface mesh depicted on the right of Figure 7.15a is not modified by the smoothing algorithms applied. Fixing almost 50% of the nodes imposes an additional restriction in smoothing. In practice, such restrictions are, for example, required if different meshes share a fixed interface surface mesh or boundary values of finite element computations are only available for prescribed positions.

The second example is the stub axle model depicted in Figure 7.15b for which a triangular surface mesh was already considered in Section 6.3.3. The STEP file of the underlying geometry model is provided courtesy of INPG by the AIM@SHAPE Mesh Repository [1]. The model of genus 17 consists of 249 faces bounded by spline curves. By setting a curvature-dependent edge length, an anisotropic volume mesh of 74,811 nodes and 400,128 tetrahedra was generated. As in the case of the modular endoprosthesis model, the triangular

(a) Modular endoprosthesis   (b) Stub axle model

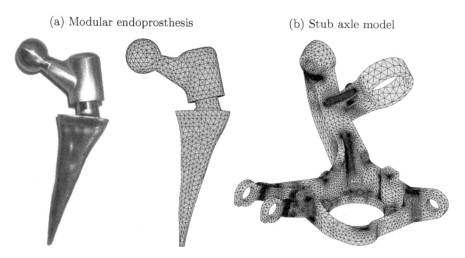

Figure 7.15: Models for tetrahedral volume meshes.

surface mesh depicted in Figure 7.15b consisting of 21,509 boundary nodes has been kept fixed during smoothing.

Tetrahedral meshes are flexible but can have an adverse topological configuration if the number of incident elements for one node becomes very large. In the case of the modular endoprosthesis mesh, 26.8 elements are attached on average to an inner node with a maximal number of 56 attached elements. In the case of the stub axle mesh, these numbers amount to 24.7 and 60, respectively. Topological adverse configurations, like many attached elements, imply that sliver tetrahedra occur, resulting in a low minimal element quality number. As in the case of the polygonal meshes, the initial tetrahedral meshes have been distorted while preserving their validity to further decrease their quality numbers.

All meshes have been smoothed by smart Laplacian smoothing, the feasible Newton-based global optimization approach of the Mesh Quality Improvement Toolkit and GETMe smoothing. In the case of the latter, the opposite normal-based regularizing transformation according to Definition 5.8 has been applied using tetrahedron volume-preserving scaling. In the case of the simultaneous and sequential substeps, the fixed transformation parameters $\sigma = 40$ and $\sigma = 0.0001$ have been applied, respectively. Here, the low σ value of the sequential substep reflects a more conservative smoothing strategy due to the adverse topological configurations. In contrast, the aggressive transformation parameter for the simultaneous substep is considered by setting the relaxation parameters of the first substep to $\varrho = 0.1$ and to $\varrho = 0.75$ for the sequential substep.

Cross sections of the modular endoprosthesis meshes are depicted in Figure 7.16. As can be seen, the initial as well as the smart Laplacian smoothing mesh contain a considerable number of very low-quality elements. Due to the partially adverse topological configuration, such elements also exist in the meshes obtained by global optimization and GETMe smoothing. However, the number of such elements is significantly decreased. For example, the initial mesh contains 2,068 elements with quality number below 0.23. The numbers of such elements after smart Laplacian smoothing, global optimization, and GETMe smoothing amount to 1,270, 38, and 34, respectively. In the case of GETMe smoothing, all these 34 tetrahedra consist only of boundary nodes. Since these are kept fixed, such elements are not improvable.

This is reflected by the resulting quality numbers given in Table 7.6. Here, q_{min} denotes the minimal element quality number and q^*_{min} the minimal quality number of all elements with at least one interior node, hence the lowest quality number of all elements, which can be improved by the smoothing methods. As can be seen, all methods improve q_{min} only slightly. However, in the case of global optimization and GETMe smoothing, this lowest element quality number is caused by a fixed element, whereas the lowest element quality number in the case of smart Laplacian smoothing occurs for an improvable element. The lowest improvable element quality number is increased by GETMe smoothing to a value of 0.2352, which is 52% larger than that obtained by global optimization.

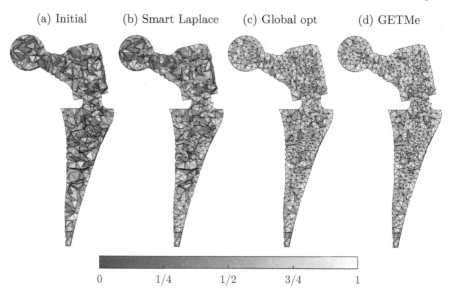

| (a) Initial | (b) Smart Laplace | (c) Global opt | (d) GETMe |

0   1/4   1/2   3/4   1

Figure 7.16: Modular endoprosthesis mesh cross sections. Tetrahedral elements are shaded according to their mean ratio quality number.

In contrast to the other smoothing methods, smart Laplacian smoothing also fails to improve the mean mesh quality significantly.

Table 7.6: Modular endoprosthesis volume mesh smoothing results.

Method	Iter	Time (s)	q_{min}	q^*_{min}	q_{mean}
Initial	–	–	0.0010	0.0010	0.4748
Smart Laplace	10	0.14	0.0022	0.0022	0.5639
Global optimization	42	0.41	0.0253	0.1544	0.7319
GETMe	77 / 200,600	0.66	0.0253	0.2352	0.7412

Due to the conservative choice of parameters, the number of 77 GETMe simultaneous iterations taking 0.65s in total and 200,600 sequential iterations applied within 0.01s is comparably high. However, as is depicted in Figure 7.17, there is a steep ascent in mean mesh quality within the first smoothing steps. After that, q_{mean} is further improved until the termination criterion of the mean mesh quality improvement achieved by one smoothing step dropping below 10^{-4} is met. Therefore, a termination criterion based on the improvement ratio could be applied alternatively to avoid such additional iterations. Furthermore, in practice larger mean mesh quality improvement thresholds can be used, since further improvements usually have no relevant impact on finite element solution accuracy. The total smoothing time of 0.01s of GETMe sequential smoothing also shows that this substep is very efficient, since only one element is transformed per iteration.

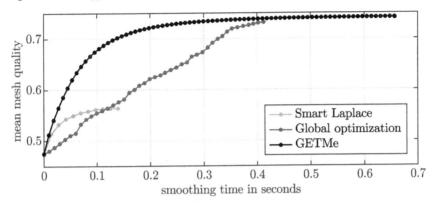

Figure 7.17: Modular endoprosthesis mean mesh quality q_{mean} with respect to smoothing time.

Figure 7.18 depicts a cross section of the upper part of the initial stub axle mesh and its counterparts after applying different smoothing algorithms. As in the case of the modular endoprosthesis mesh, very low-quality elements consisting entirely of boundary nodes exist. Since these boundary nodes are kept fixed, the minimal element quality cannot be improved beyond $q_{min} = 0.0254$ in the given example, without incorporating topological modifications or surface mesh smoothing techniques into the mesh smoothing process. However, by applying such algorithms, results would not be comparable with those of the global optimization-based approach. Therefore, only inner mesh nodes have been improved. The resulting mesh quality numbers and smoothing times in seconds are given in Table 7.7.

Table 7.7: Stub axle volume mesh smoothing results.

Method	Iter	Time (s)	q_{min}	q^*_{min}	q_{mean}
Initial	–	–	0.0001	0.0001	0.5284
Smart Laplace	11	8.63	0.0002	0.0002	0.7480
Global optimization	54	20.52	0.0021	0.0021	0.7489
GETMe	64 / 810,700	30.55	0.0254	0.2664	0.7789

In contrast to the other smoothing methods, GETMe increases the minimal element quality number to the optimal value. Furthermore, the minimal element quality number of improvable elements is increased to $q^*_{min} = 0.2664$, which is significantly better than those of smart Laplacian smoothing and global optimization, for which q_{min} and q^*_{min} stagnate on a low level. The number of tetrahedra with quality numbers below 0.2664 amount to 57,288, 7,354, 1,584, and 102 in the case of the initial mesh and its counterparts after applying smart Laplacian smoothing, global optimization, and GETMe, respectively.

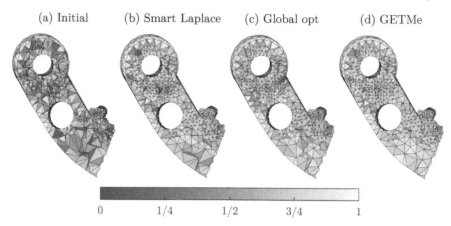

Figure 7.18: Stub axle mesh cross sections. Tetrahedral elements are shaded according to their mean ratio quality number.

This is also visible in the low quality elements shaded dark gray in the cross sections depicted in Figure 7.18.

Since GETMe smoothing can further improve the mesh quality numbers, its iteration numbers and the resulting runtime is increased if compared to the other methods. Nevertheless, as in the case of the other examples, GETMe smoothing shows a favorable runtime behavior by a significant improvement of the mean mesh quality within the first smoothing steps, as is depicted in Figure 7.19. In total, 438 elements had to be reverted during the simultaneous GETMe substep. This phase of GETMe smoothing took 30.5s seconds and resulted in the quality numbers $q^*_{min} = 0.0019$ and $q_{mean} = 0.7791$. The subsequent 810,700 sequential substeps took 0.05s in total and led to a slight decrease of the mean mesh quality to a final value of $q_{mean} = 0.7789$ while increasing the minimal improvable element quality to $q^*_{min} = 0.2664$. During smart Laplacian smoothing, 7,217 elements had to be reverted to their previous node positions to avoid the generation of invalid elements.

As in the previous examples, the feasible Newton approach of the global optimization-based smoothing method was terminated by its own inverse mean ratio-based criterion. For comparison reasons, this approach was alternatively terminated by the same criterion as the other two methods; that is, if the mean quality number difference of two consecutive iterations dropped below 10^{-4}. This resulted in 79 iterations taking 29.3s in total. In doing so, the mean mesh quality number was further improved to 0.7770, which is comparable to the result obtained by GETMe smoothing. In contrast, the resulting minimal element quality number $q^*_{min} = 0.0269$ does not differ significantly from that given in Table 7.7.

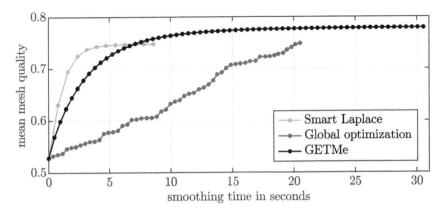

Figure 7.19: Stub axle mean mesh quality q_{mean} with respect to smoothing time.

7.2.2 Hexahedral meshes

In specific finite element and finite volume applications, such as, for example, elastic and elastic-plastic solid continuum analysis, hexahedral meshes are preferred, since more suitable trial functions can be defined on such tessellations, resulting in a higher solution accuracy combined with a lower number of mesh elements [10, 24]. However, depending on the geometrical complexity of the simulation domain, the generation of good-quality all-hexahedral meshes is in general considerably harder than the generation of tetrahedral meshes. In the case of simple domains, straightforward approaches like sweeping, i.e., extruding a quadrilateral surface mesh, or template-based refinement techniques can be applied. This section first discusses the results for two prototypical meshes based on such generation techniques.

The first is the structured hexahedral mesh depicted in Figure 7.20a. It was obtained by defining the nodes as the Cartesian product of 13 non-equidistant x values, four equidistant y-values, and 11 non-equidistant z values. In x direction, the grid widths vary from 0.005 to 0.5, that is, by factor 100. In z direction they vary from 0.005 to 1.0, that is, by factor 200. The regular grid width in y direction is given by 2.0. The resulting mesh quality numbers range from 0.0010 to 0.5714 with a mean mesh quality of 0.0943. The mesh consists of 572 nodes and 360 hexahedral elements. Since the mesh is structured, each of its 198 interior nodes is shared by 8 incident elements. This mesh has not been distorted.

The second mesh was obtained by iteratively refining an initial cube of edge length 500 ten times towards its center. A cross section of the resulting undistorted mesh is depicted in Figure 7.20b. Here, mesh refinement is based on the octree meshing templates described in [76]. To be precise, the templates T_2, T_4, and T_8 depicted in Figure 7.21 have been applied to the eight cubes obtained by halving the cube in each coordinate direction according to the

(a) Nonuniform mesh   (b) Refined mesh

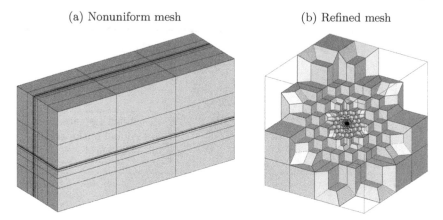

Figure 7.20: Nonuniform hexahedral grid mesh and a cross section of a mesh generated by template-based refinement.

(a) T_2 (b) T_4 (c) T_8

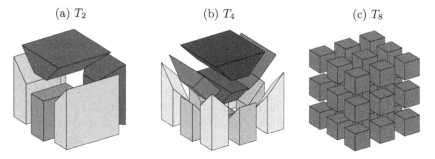

Figure 7.21: Refinement templates for hexahedral elements. Different element shapes per template are indicated by different shades of gray.

scheme

$$\text{upper layer:} \quad \frac{T_2 \mid \text{none}}{T_4 \mid T_2} \qquad \text{lower layer:} \quad \frac{T_4 \mid T_2}{T_8 \mid T_4},$$

where each cell indicates the refinement template to be applied. After that, the eight elements forming the lower outer cube of the template T_8 have been iteratively subdivided following the same refinement scheme. The final mesh has been obtained by assembling eight of these refined meshes as octants, resulting in 6,165 nodes and 5,984 elements in total.

Due to this iterative refinement process, the resulting mesh consists of scaled versions of the hexahedra depicted in the exploded template meshes shown in Figure 7.21. Here, for each template, elements with the same shape are marked by the same shade of gray. The mean quality ratio numbers of the hexahedra of T_2 depicted in Figure 7.21a are given by 0.6357, 0.7048, and 0.8399, respectively. Those of T_4 depicted in Figure 7.21b range from 0.5400 to 1.0. Finally, the

27 elements of T_8 depicted in Figure 7.21c represent regular hexahedra with the optimal mean ratio quality number 1.0. Due to construction, these are also the element quality numbers of the initial undistorted mesh depicted in Figure 7.20b obtained by iterative refinement. This mesh has additionally been distorted by element validity-preserving random node movements to further decrease its quality.

The nonuniform grid mesh as well as the distorted refined mesh have been smoothed by smart Laplacian smoothing, global optimization, and GETMe smoothing. In the case of the latter, elements have been transformed using the dual element-based approach according to Definition 5.9 using a quality dependent transformation parameter $\sigma \in [0.5, 0.6]$, the relaxation parameter $\varrho = 3/4$, the exponent $\eta = 1/2$, and edge length-preserving element scaling in the simultaneous substep. For the sequential substep, the fixed transformation parameter $\sigma = 0.01$ and the penalty correction parameters $\Delta \pi_i = 2 \cdot 10^{-4}$, $\Delta \pi_r = 4 \cdot 10^{-4}$, and $\Delta \pi_s = 2 \cdot 10^{-4}$ have been applied.

Resulting smoothing iteration numbers, the total smoothing time, and the mesh quality numbers for both examples and different smoothing methods are given in Table 7.8. Due to the mesh construction schemes applied, each element has at least one interior node. Hence, for the two examples given it holds that $q_{min} = q^*_{min}$. In the case of GETMe smoothing, the iteration numbers of the simultaneous and the sequential substep are given. Since mesh quality is evaluated after each 100th cycle in the sequential substep and termination is based on 10 cycles without q_{min} improvement, the second iteration numbers are always multiples of 100.

Table 7.8: Smoothing results for the nonuniform and the template-based refined hexahedral mesh.

Mesh	Method	Iter	Time (s)	q_{min}	q_{mean}
Nonuniform	Initial	–	–	0.0010	0.0943
	Smart Laplace	13	0.03	0.0053	0.1080
	Global opt	25	0.04	0.0054	0.1199
	GETMe	22 / 1,000	0.02	0.0209	0.1195
Refined	Initial	–	–	0.1370	0.4979
	Smart Laplace	14	0.57	0.1721	0.7905
	Global opt	37	1.25	0.5152	0.8336
	GETMe	14 / 1,600	0.13	0.6603	0.8251

Cross sections of the initial and smoothed nonuniform hexahedral meshes are depicted in Figure 7.22. For a better comparability, all subfigures show the same elements. That is, cross sections are not obtained by removing the elements on one side of a clipping plane for each mesh individually. Rather, the elements to be shown have been determined by clipping the initial mesh by a clipping plane.

The initial mesh depicted in Figure 7.22a shows the regular but nonuniform

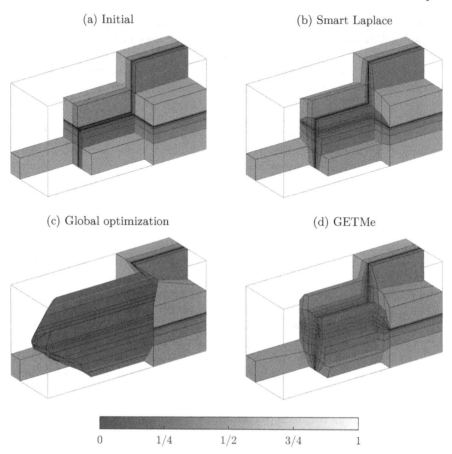

Figure 7.22: Nonuniform hexahedral grid mesh cross sections with elements shaded according to their mean ratio quality number.

structure of the mesh and the strongly varying element quality due to the substantial difference in the dimensions of the elements. The result obtained by smart Laplacian smoothing depicted in Figure 7.22b shows that thin layers are moderately thickened after smoothing, increasing the initial minimal element quality from 0.0010 to 0.0053 and the initial mean mesh quality from 0.0943 to 0.1080. In contrast, the global optimization-based smoothing approach leads to unnaturally stretched inner elements with respect to the y dimension as can be seen in Figure 7.22c. This is because the global optimization approach entirely focuses on improving the mean mesh quality. For the given mesh, the resulting quality numbers are $q_{min} = 0.0054$ and $q_{mean} = 0.1199$. GETMe smoothing significantly improves the minimal element quality to 0.0209 and the mean mesh quality to 0.1195. As can be seen in Figure 7.22d, the uniform mesh width with respect to the y direction is almost preserved. However, with

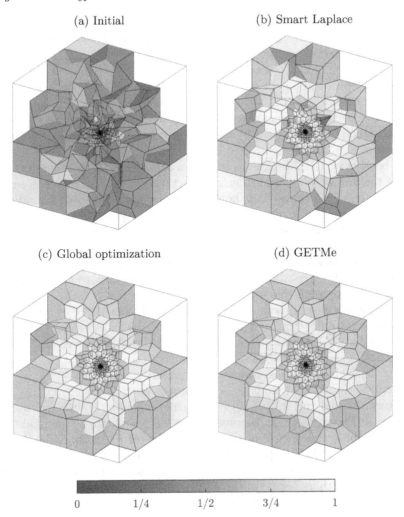

Figure 7.23: Refined hexahedral mesh cross sections with elements shaded according to their mean ratio quality number.

respect to the other dimensions, the inner element sizes are more balanced if compared to the results of smart Laplacian smoothing.

Figure 7.23 depicts cross sections for the refined and distorted mesh and their smoothed counterparts. By element validity-preserving node distortions, the initial mesh quality numbers have been decreased to $q_{min} = 0.1370$ and $q_{mean} = 0.4979$. This is reflected by the large number of low quality elements depicted in Figure 7.23a and the less uniform shape of the elements if compared to the undistorted version depicted in Figure 7.20b. As in the previous examples, smart Laplacian smoothing failed to improve the minimal element quality to the same extent as the other methods. This also holds for the mean mesh quality, which is also due to the remaining lower quality elements. For example, the number of elements below the threshold $q_{min} = 0.6603$ obtained by GETMe smoothing amounts to 1,415 and 246 in the case of smart Laplacian smoothing and global optimization. In contrast to smart Laplacian smoothing, global optimization as well as GETMe smoothing almost restore the ideal hexahedral elements in each refinement layer. The slightly lower mean mesh quality of GETMe smoothing is caused by further improving q_{min}, leading to a slight decrease of neighboring elements quality.

In both cases, GETMe again shows a favorable steep ascent in mesh quality during the first iterations. Furthermore, iteration numbers are low, resulting in short smoothing times. This holds particularly for hexahedral meshes as has been observed in various other smoothing examples. The reason for this is on the one hand the strong regularizing effect of the dual elements-based hexahedron transformation, as indicated by Figure 5.24. On the other hand, the topological configuration of hexahedral meshes is usually more favorable if compared to other mesh types such as tetrahedral meshes, which additionally support effective mesh smoothing.

The last hexahedral mesh example considers the inner core of a real-world computational fluid dynamics mesh of the car Aletis developed by TWT GmbH Science & Innovation, commissioned by the Hellenic Vehicle Industry ΕΛΒΟ S.A. Thessaloniki [138]. The car was presented in 2001 at the IAA, the International Motor Show Germany in Frankfurt am Main.

Pictures of the fully functional Aletis prototype are shown in Figure 7.24a. A feature-reduced surface mesh of the model including a driver dummy is depicted in Figure 7.24b. In the underlying CFD simulation, the model was assumed to be symmetric. Hence only half of the car has been modeled and a hexahedral mesh of the exterior of the car has been generated. Figure 7.24c depicts the rotated part of the volume mesh, which is considered in the subsequent smoothing tests. For the CFD simulation, this mesh is extended by a hexahedral interface mesh forming the transition to an all-regular hexahedral mesh of the flow passage. Since the flow passage mesh as well as the transition mesh are of high quality, they have been removed in the given example.

The remaining mesh considered consists of 1,710,978 nodes and 1,610,234 hexahedral elements. All 201,369 boundary nodes have been kept fixed during smoothing to preserve the surface mesh of the car model and the compatibility with the interface mesh. Due to the CFD-focused meshing approach, elongated hexahedral elements are used near the surface of the car, incorporating a specific turbulence model. In practice, such turbulence layers, usually generated by advancing front techniques, are either kept fixed during smoothing or handled similar to the boundary layer mesh example described in Section 7.1.3.1 using a layer-specific element transformation or scaling scheme. However, to demonstrate the smoothing potential of the methods under consideration, these layers have also been smoothed, which bounds the achievable mean mesh quality from above. The initial mesh has additionally been distorted by mesh

(a) Fully functional Aletis prototype

(b) Aletis model surface mesh

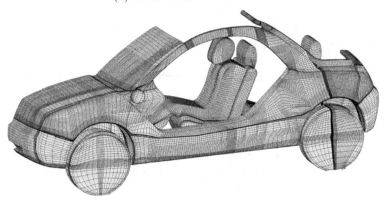

(c) Exterior hexahedral volume mesh

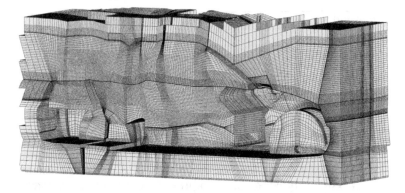

Figure 7.24: Fully functional Aletis prototype and its meshes.

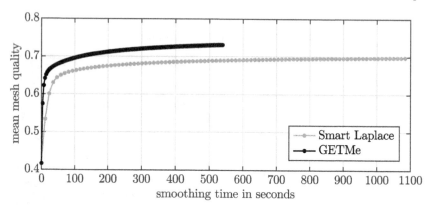

Figure 7.25: Aletis mean mesh quality q_{mean} with respect to smoothing time.

validity-preserving random node movements resulting in the quality numbers $q_{\mathrm{min}} = 0.0042$ and $q_{\mathrm{mean}} = 0.4165$. This distorted initial mesh has been smoothed by smart Laplacian smoothing, global optimization, and GETMe smoothing using the same parameters as in the case of the previous hexahedral meshes.

The resulting smoothing times and quality numbers are given in Table 7.9. Smart Laplacian smoothing terminated after 94 iterations, each taking 11.53 seconds on average. Although the initial mesh is valid, the global optimization approach of Mesquite generated 15 inverted elements during smoothing and could not proceed due to invalid gradient computations. The GETMe simultaneous substep terminated after 166 iterations, each taking 3.24 seconds on average. The subsequent GETMe sequential substep terminated after 3,000 iterations taking 0.03 seconds in total.

Table 7.9: Aletis mesh quality numbers and smoothing times.

Method	Iter	Time (s)	q_{min}	q_{mean}
Initial	–	–	0.0042	0.4165
Smart Laplace	94	1,083.87	0.0045	0.6978
Global optimization	failed	–	–	–
GETMe	166 / 3,000	538.68	0.0062	0.7320

In both cases, the achieved minimal element quality is very low, due to the CFD-focused density of the mesh. In contrast, the mean mesh quality is improved significantly in both cases. As in the case of the previous meshes, the first iterations of both smoothing methods lead to a steep ascent in mean mesh quality. For example, q_{mean} is improved by 0.1842 and 0.2062 in the first two iterations of smart Laplacian and GETMe simultaneous smoothing, respectively. After this steep ascent, GETMe smoothing continuously improves the mean mesh quality. In contrast, the improvement graph of q_{mean} with

(a) Initial

(b) Smart Laplace

(c) GETMe

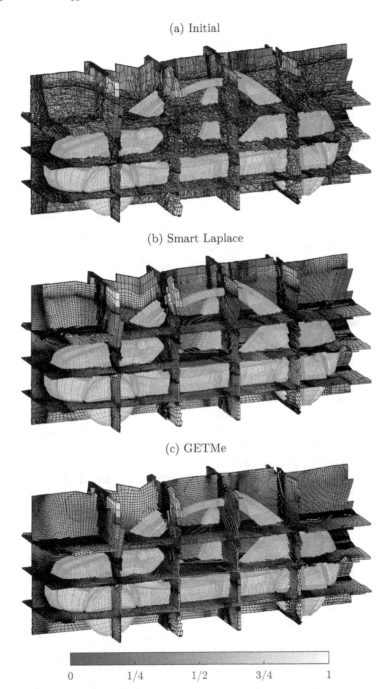

0	1/4	1/2	3/4	1

Figure 7.26: Aletis mesh slices with elements shaded according to their mean ratio quality number.

respect to smoothing time achieved by smart Laplacian smoothing only shows slight increases in the second half of the smoothing process, as is shown by Figure 7.25.

Figure 7.26 depicts slices of the initial volume mesh and its smoothed versions. As can be seen by the quality-shaded elements, the initial mesh is severely distorted. For example, the number of elements with a quality number below 0.1 amounts to 37,804. Smart Laplacian smoothing reduces this number to 18,780, which is about 49.7% of the initial value. However, this substantial number of low quality elements is visible in the mesh slices depicted in Figure 7.26b by the elements marked with dark gray shading. After GETMe smoothing, 2,711 elements with quality number below 0.1 remain, which is 7.2% of the initial value. In doing so, GETMe smoothing is about two times faster if compared with smart Laplacian smoothing. The resulting mesh quality is also visible from the elements with light gray shading in Figure 7.26c. However, low quality elements remain in regions with disadvantageous element dimension ratios due to the CFD-focused meshing approach.

7.3   Mixed element type volumetric meshes

7.3.1   Embedded meshes

All volume meshing techniques have specific advantages and drawbacks. For example, simple parameterized models or constructive solid geometry models can easily be meshed into hexahedral elements by sweeping or advancing front techniques. Depending on the application, hexahedral elements are more suitable for finite element computations and thus are preferred. However, there is yet no general and numerically robust approach in hexahedral meshing of arbitrarily complex volume models as provided by Delaunay triangulation for tetrahedral meshing. Instead of using single type volume meshes, each subdomain of the simulation model can alternatively be meshed by a suitable technique, and the resulting meshes can be glued together by transition elements based, for example, on pyramidal and prismatic elements. Such meshes are discussed in the following.

Two different models will be considered in which tetrahedral and hexahedral meshes are connected by pyramidal elements. The first is an embedding of a triangular surface mesh of a ring depicted in Figure 7.27a. It consists of 5,226 triangular elements and is provided by the Gamma project mesh database [56]. This mesh is embedded into a hexahedral mesh of a cuboid originally consisting of $24 \times 25 \times 16$ regular elements. Three layers of hexahedra have been removed around the ring mesh. On each face of the resulting interior quadrilateral surface mesh, a pyramidal element was placed pointing towards the ring. In doing so, an interior triangular interface surface mesh was generated. This and

(a) Ring surface mesh　　　　　　(b) Ring embedding mesh

(c) Plate volume mesh　　　　　　(d) Plate embedding mesh

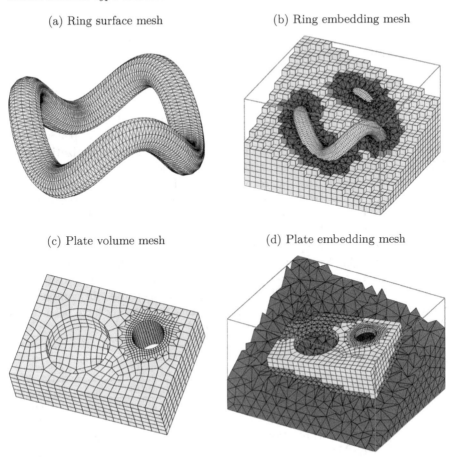

Figure 7.27: Embedded model meshes and clipped surrounding meshes with elements shaded by element type.

the triangular surface mesh of the ring model are bounding a volume, which was tessellated into tetrahedral elements by Delaunay meshing. That is, the ring is embedded into a tetrahedral mesh connected to one layer of pyramidal elements, which build the interface to a regular hexahedral mesh. A clipped version of this mesh, consisting of 74,106 tetrahedra marked dark gray, 1,618 pyramids marked medium gray, and 7,486 hexahedra marked light gray, is depicted in Figure 7.27b. The depicted mesh has additionally been distorted by element validity-preserving random node movements to decrease the initial mesh quality.

The second example considers the hexahedral sweep mesh of a plate with two drill holes depicted in Figure 7.27c and its embedding into a tetrahedral mesh of a cuboid. The hexahedral plate mesh consists of 5,748 elements. It is refined towards the drill hole with smaller diameter. By placing a pyramidal element on each of the 2,220 quadrilateral surface mesh elements, a triangular interface was generated. This wrapped mesh was placed inside a bounding cuboid with triangular surface mesh. The volume bounded by the triangular meshes of the cuboid and the wrapped plate was filled by 26,082 tetrahedra using a Delaunay-based meshing approach. A clipped version of the resulting mesh is depicted in Figure 7.27d. Again, tetrahedral, pyramidal, and hexahedral elements are marked with dark, medium, and light gray shading, respectively.

The quality numbers of the initial meshes are given in Table 7.10 along with the results obtained by smart Laplacian smoothing, global optimization, and GETMe smoothing. As in the previous section, the mean mesh quality-based termination tolerance has been set to 10^{-4}. In the case of GETMe smoothing, the dual element-based element transformations according to Definition 5.9 with element type-specific and element quality-controlled transformation parameters $\sigma^{\mathrm{tet}} \in [0.77, 0.84]$, $\sigma^{\mathrm{hex}} \in [2.57, 3.45]$, and $\sigma^{\mathrm{pyr}} = 1.86$ have been used in the simultaneous substep. The relaxation parameter was set to $\varrho = 2/3$, the quality weight exponent to $\eta = 1/4$, and edge length-preserving scaling was applied. For the sequential substep, the fixed transformation parameters $\sigma^{\mathrm{tet}} = 0.81$, $\sigma^{\mathrm{hex}} = 2.74$, and $\sigma^{\mathrm{pyr}} = 1.82$ have been used in combination with the relaxation parameter $\varrho = 0.01$. Penalty parameters have been set to $\Delta\pi_i = 10^{-2}$, $\Delta\pi_r = 5 \cdot 10^{-4}$, and $\Delta\pi_s = 10^{-2}$.

Table 7.10: Smoothing results for embedded meshes.

Mesh	Method	Iter	Time (s)	$q_{\min}$	q_{mean}
Ring	Initial	–	–	0.0005	0.4499
	Smart Laplace	15	3.30	0.0005	0.6542
	Global optimization	40	4.26	0.0276	0.7275
	GETMe	21 / 6,800	2.72	0.2330	0.7435
Plate	Initial	–	–	0.1625	0.7302
	Smart Laplace	7	0.67	0.0258	0.8193
	Global optimization	16	1.39	0.3305	0.8385
	GETMe	13 / 10,600	0.77	0.4446	0.8299

In the case of the embedded ring mesh, 11,285 boundary nodes located on the ring as well as the cuboid surface have been kept fixed during smoothing in order to preserve the shape of the model. The remaining 14,246 nodes have been moved by the smoothing algorithms to improve mesh quality. None of the elements consist of boundary nodes only, hence $q_{\min}$ equals $q^*_{\min}$ in all cases. Due to mesh distortion, the initial minimal element quality given by 0.0005 is very low, as is the mean mesh quality, resulting in 0.4499. The number of elements with a mean ratio number below 0.2330 amounts to 12,922. This is also visible in the initial mesh cross section depicted in Figure 7.28a. Although smart Laplacian smoothing fails to improve $q_{\min}$, the mean mesh quality is improved by about 45% within 15 iterations. During smoothing, 2,057 invalid

(a) Initial

(b) Smart Laplace

(c) Global optimization

(d) GETMe

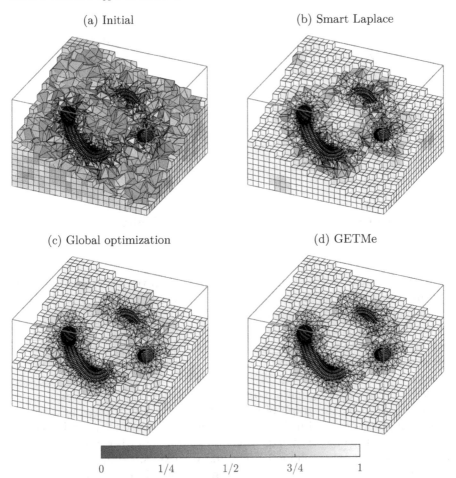

0   1/4   1/2   3/4   1

Figure 7.28: Embedded ring mesh cross sections with elements shaded according to their mean ratio quality number.

elements had to be reset in order preserve mesh validity. The final number of elements with quality number below 0.2330 amounts to 3,364.

Due to its mean mesh quality-focused approach, global optimization-based smoothing fails to improve q_{min} significantly although the number of low quality elements is considerably reduced. For example, the number of elements with $q(E) < 0.2330$ is decreased to 218. In contrast, the mean mesh quality number is improved by about 62% if compared to the initial mesh. The improved mesh quality is also visible from the resulting mesh cross section depicted in Figure 7.28c.

Although GETMe smoothing is based on geometric construction schemes, its resulting mesh quality is remarkable. Due to the combined approach, there

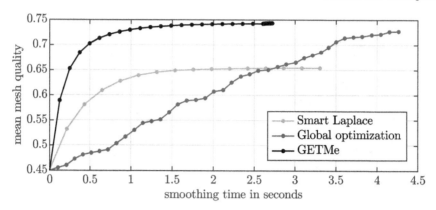

Figure 7.29: Embedded ring mean mesh quality q_{mean} with respect to smoothing time.

is no element with $q(E) < 0.2330$ and q_{mean} is improved by 65%. That is, in the given example, the mean mesh quality is even better than that obtained by the global optimization-based approach.

Smart Laplacian smoothing is based on element type independent neighboring node averaging and global optimization on node relocation based on mean mesh quality. In contrast, by the dual element-based transformation approach, GETMe smoothing tries to regularize each element individually, as is visible in the resulting mesh cross section depicted in Figure 7.28d.

As is shown by the graph of the mean mesh quality with respect to the smoothing time depicted in Figure 7.29, the performance of smart Laplacian smoothing in the case of mixed volume meshes is lower if compared to that of single element-type volume meshes and GETMe smoothing. Global optimization leads to significant improvements, although it is comparably slow. For example, global optimization takes about seven times longer to achieve $q_{\mathrm{mean}} = 0.7$ if compared to GETMe smoothing. A quality threshold that cannot be achieved by smart Laplacian smoothing, but is surpassed by GETMe simultaneous smoothing after only four iteration steps.

The embedded plate model mesh has not been distorted to demonstrate further aspects of mesh smoothing by the methods under consideration. The mesh consists of 34,050 elements in total and 11,371 nodes, of which 1,022 are boundary nodes on the outer cuboid faces kept fixed during smoothing. In contrast, the surface nodes of the plate model have not been kept fixed and no constraints have been used to allow an unhampered smoothing of adjacent elements of different types. Due to not distorting the mesh, the initial quality numbers already amount to $q_{\mathrm{min}} = 0.1625$ and $q_{\mathrm{mean}} = 0.7302$, as is also reflected by the cross section depicted in Figure 7.30a. Here, low quality elements are in the transition layer consisting of pyramidal and the attached tetrahedral elements.

For the given example, the smoothing results given in the second half of

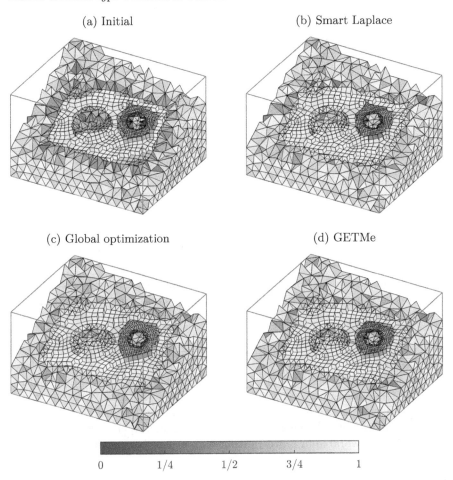

Figure 7.30: Embedded plate mesh cross sections with elements shaded according to their mean ratio quality number.

Table 7.10 demonstrate a drawback of smart Laplacian smoothing. The minimal element quality $q_{min} = 0.1625$ of the initial mesh is significantly decreased to 0.0258 by smart Laplacian smoothing, which is problematic for subsequent finite element computations. In contrast q_{min} is significantly improved by global optimization and GETMe smoothing. For example, the numbers of elements with $q(E) < 0.4446$ amount to 2,240, 1,190, 120, and 0 in the case of the initial mesh and those obtained by smart Laplacian smoothing, global optimization, and GETMe smoothing, respectively. The mean mesh quality improvement by these methods amounts to about 12%, 15%, and 14%, respectively.

Resulting mean mesh quality numbers with respect to the smoothing time for the embedded plate mesh are depicted in Figure 7.31. As can be seen, although the initial mesh is not distorted, quality improvements are only

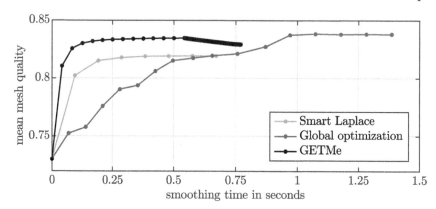

Figure 7.31: Embedded plate mean mesh quality q_{mean} with respect to smoothing time.

moderate for each iteration of the global optimization-based approach. For example, if compared to GETMe simultaneous smoothing, seven iterations of global optimization are required to reach the result obtained by the first GETMe simultaneous iteration.

In the case of GETMe smoothing, the graph shows a slight decrease in mean mesh quality during the GETMe sequential substep. This is caused by the fact that improving low quality elements affects the quality of neighboring elements, resulting in a slight decrease of mean mesh quality in the given example. However, if compared to global optimization, the mean mesh quality is only decreased by 1%, whereas the minimal element quality is improved by 35%, which is favorable in finite element-focused mesh smoothing.

7.3.2   Layered mesh

In the following, a tetrahedral mesh of an aorta wrapped with layers of prismatic elements will be considered. Such mixed meshes are used, for example, in computational biofluid dynamics applications [38]. The model, as well as its initial tetrahedral mesh, is provided courtesy of MB-AWG by the AIM@SHAPE Mesh Repository [1]. It consists of 35,551 tetrahedra, which have been wrapped with six layers of prisms with a constant height of 0.012. Here, the height was chosen small to avoid the intersection of layer elements as well as to avoid changing the model surface too much.

Figure 7.32a depicts a clipped version of the resulting layered mesh with tetrahedral elements shaded dark gray and prismatic elements shaded light gray. The part of the aorta under consideration is marked by a black rectangle in the illustration depicted in Figure 7.32b.

The complete mesh consists of 49,903 nodes with 6,729 boundary nodes kept fixed during smoothing. The total number of mesh elements amounts to 116,275. For smoothing, the same termination and control parameters

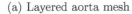

(a) Layered aorta mesh   (b) Aorta illustration

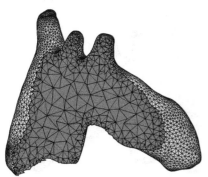 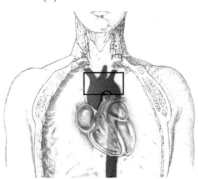

Figure 7.32: Clipped layered aorta mesh with elements shaded by element type and illustration of the aorta part under consideration. Illustration provided by J. Heuser, Wikimedia Commons, license CC-BY-SA-3.0, GFDL [66].

have been used as described in the previous section. In the case of GETMe simultaneous smoothing, the dual element-based transformation scheme with the element type-specific transformation parameters $\sigma^{\mathrm{tet}} \in [0.77, 0.84]$ and $\sigma^{\mathrm{pri}} = 1.58$ have been applied. In the case of the sequential smoothing substep, transformation parameters have been set to $\sigma^{\mathrm{tet}} = 0.81$ and $\sigma^{\mathrm{pri}} = 0.85$. The initial mesh quality numbers, as well as those obtained by applying various smoothing methods, are given in Table 7.11.

Table 7.11: Smoothing results for the layered aorta mesh.

Method	Iter	Time (s)	$q_{\min}$	q_{mean}
Initial	–	–	0.0413	0.3077
Smart Laplace	58	29.06	0.0064	0.5724
Global optimization	127	52.29	0.2176	0.5956
GETMe	182 / 4,000	33.11	0.2611	0.8226

As can be seen, the minimal element quality number $q_{\min} = 0.0413$ as well as the mean mesh quality number $q_{\mathrm{mean}} = 0.3077$ of the initial mixed mesh are low. Here, the initial tetrahedral core mesh with $q_{\min} = 0.2677$ and $q_{\mathrm{mean}} = 0.8098$ is already of good quality. In contrast, due to the thin elements, the initial prismatic layer mesh is of very low quality as can be seen by its quality numbers, resulting in $q_{\min} = 0.0413$ and $q_{\mathrm{mean}} = 0.0866$. In consequence, mesh improvement can mainly be achieved by improving the prismatic layers, which also affects the attached tetrahedral core mesh. To demonstrate the characteristics of the smoothing methods under consideration, no constraints for the prismatic layer heights have been incorporated. In practice, such constraints would have been incorporated to consider requirements by specific

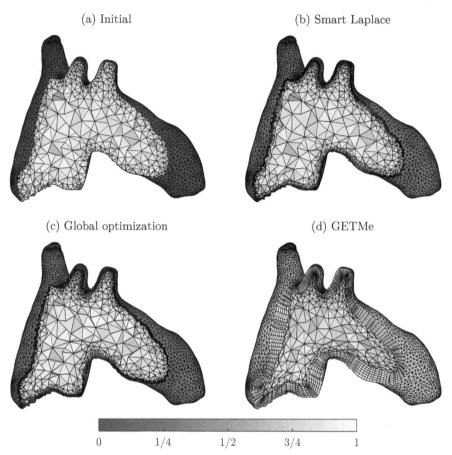

Figure 7.33: Clipped layered aorta meshes with elements shaded according to their mean ratio quality number.

turbulence models near the mesh boundary. Furthermore, in contrast to some previous examples, the initial mesh has not been distorted.

The high quality tetrahedral mesh, as well as the low quality prismatic layers, are visible in the clipped initial mesh depicted in Figure 7.33a. Here, elements are shaded according to their mean ratio quality number, with dark gray indicating low quality elements and light gray indicating high quality elements. Figure 7.33b depicts the results obtained by smart Laplacian smoothing. As can be seen, the prismatic layers are slightly thickened, which improves element quality numbers. However, in contrast to the initial mesh, the minimal element quality number is decreased to $q_{min} = 0.0064$, whereas the mean mesh quality number is improved to $q_{mean} = 0.5724$. Here, the minimal and mean mesh quality numbers for the tetrahedral core mesh are given by $q_{min} = 0.0064$ and

$q_{\mathrm{mean}} = 0.7872$. The quality numbers of the prismatic layer elements obtained by smart Laplacian smoothing are $q_{\mathrm{min}} = 0.0448$ and $q_{\mathrm{mean}} = 0.4778$.

For the given example, the global optimization-based approach did significantly improve the minimal element quality number to $q_{\mathrm{min}} = 0.2176$. Since the minimal tetrahedral element quality is given by $q_{\mathrm{min}} = 0.3451$, the lowest quality number element is a prism. Compared to smart Laplacian smoothing, the overall mean mesh quality number $q_{\mathrm{mean}} = 0.5956$ is further improved but still not on a high level. Here, the mean mesh quality numbers for the tetrahedral core mesh and the prismatic layer mesh amount to $q_{\mathrm{mean}} = 0.8085$ and $q_{\mathrm{mean}} = 0.5018$, respectively. By comparing these numbers with the initial mesh quality numbers, it can be seen that improving the prismatic layer mesh does not come at the cost of decreasing the tetrahedral core mesh quality considerably. The improved prismatic elements are also visible from the clipped mesh shown in Figure 7.33c.

Finally, Figure 7.33d depicts a clipped version of the result obtained by applying GETMe smoothing. As can be seen, GETMe smoothing thickens the prismatic layers to a much larger extent than the other methods. In consequence, the prismatic layer mesh is of much higher quality, with quality numbers resulting in $q_{\mathrm{min}} = 0.2611$ and $q_{\mathrm{mean}} = 0.8416$. The quality numbers of the tetrahedral core mesh are given by $q_{\mathrm{min}} = 0.3673$ and $q_{\mathrm{mean}} = 0.7797$. Here, the tetrahedral mean mesh quality is slightly decreased if compared to the initial mesh. However, the minimal tetrahedral element quality number is significantly improved, as are the quality numbers of the prismatic layer mesh. Furthermore, the geometry-driven approach of GETMe smoothing improves the mean mesh quality even more than the mean mesh quality-geared global optimization-based approach. In doing so, the minimal element quality number and mean mesh quality number achieved by GETMe smoothing is 20% and 38% better than that obtained by the global optimization-based approach.

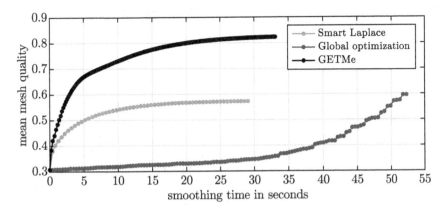

Figure 7.34: Layered aorta mean mesh quality q_{mean} with respect to smoothing time.

As in the case of the other smoothing examples, the first iterations of GETMe smoothing are very effective, as can be seen in the graph of mesh quality with respect to smoothing time depicted in Figure 7.34. That is, this is not an effect caused by effectively removing artificial mesh distortions; rather, it is an intrinsic property of the method. For example, in the given example, the final mean mesh quality number $q_{\mathrm{mean}} = 0.5956$ obtained by global optimization after 127 iterations taking 52.29 seconds in total is achieved by GETMe smoothing after only 14 iterations, taking 2.76 seconds in total, which is about 19 times faster. It can also be seen that in contrast to smart Laplacian smoothing and GETMe smoothing, significant mesh improvement by global optimization is mainly achieved during the last third of the smoothing time. Thus, in contrast to both of the other methods, a preliminary termination of the global optimization-based approach is not recommended.

7.4   Finite element method results

In this section, the influence of mesh quality on finite element solution efficiency and accuracy will be discussed in detail in the example of planar and volumetric meshes. Furthermore, the influence of different element type meshes and associated trial functions of different order for the same simulation domain will be discussed. By prescribing an exact solution, exact errors can be computed to provide an in-depth analysis of the role of mesh smoothing methods in the finite element approximation process.

7.4.1   Model problem and error norms

In the following, the *Poisson problem* with inhomogeneous boundary conditions will be considered. It is given by

$$-\Delta u = f \quad \text{in } \Omega, \tag{7.1}$$

$$u = g \quad \text{on } \partial\Omega, \tag{7.2}$$

where $\Delta = \mathrm{div}\,\mathrm{grad}$ denotes the Laplace operator, $\Omega \subset \mathbb{R}^m$ with $m \in \{2, 3\}$ the simulation domain, and $\partial\Omega$ its boundary. This model problem can be interpreted as a steady state heat conduction problem with homogeneous material and the thermal diffusivity set to one. The solution $u : \overline{\Omega} \to \mathbb{R}$ describes the temperature over the closure $\overline{\Omega}$ of Ω. The right-hand side $f : \Omega \to \mathbb{R}$ represents the heat source inside the domain, and the boundary condition $g : \partial\Omega \to \mathbb{R}$ represents the prescribed boundary temperature.

In practice, only the heat source f and the boundary temperature g are known, but not the solution u. Based on this information, an approximate solution $u_h : \overline{\Omega} \to \mathbb{R}$ of u can be computed by using the finite element approach as described in Chapter 3. To do so, the inhomogeneous model problem (7.1) is first transformed into the homogeneous Poisson problem given by Definition 3.1, as described in Section 3.2.1. After that, u_h is computed for a mesh of $\overline{\Omega}$ by the Ritz-Galerkin method described in Section 3.2.4. That is, $u_h = \sum_{k=1}^{n_h} c_k b_k$ is represented as a linear combination of weighted trial functions b_k defined over the mesh elements. The coefficients c_k are derived by solving the linear Ritz-Galerkin system $Gc = \hat{f}$ according to (3.12), where c denotes the vector of the coefficients c_k and $\hat{f}$ the right-hand side derived with the aid of f. Thus, the finite element solution process consists of the steps given by Algorithm 10.

Algorithm 10: Finite element simulation of the model problem.

   Input   : Simulation domain $\overline{\Omega}$ and functions f, g
   Output: Approximate solution u_h

1 Mesh generation for the domain $\overline{\Omega}$;
2 Mesh smoothing;
3 Assembling of the Ritz-Galerkin system $Gc = \hat{f}$;
4 Solving of the Ritz-Galerkin system;
5 Evaluation of the approximate solution $u_h = \sum_{k=1}^{n_h} c_k b_k$;

In the subsequent examples, it would be desirable to know the exact solution u to compute exact errors for the approximate solution u_h. Therefore, an exact solution u is prescribed and f is derived by applying the Laplace operator to u and interchanging signs. The boundary conditions are determined by evaluating the prescribed exact solution u along the boundary $\partial\Omega$, i.e., g is the restriction of u to $\partial\Omega$. This results in the required input data f and g for the finite element solution process, which in turn provides the approximate solution u_h. In doing so, not only the finite element solution u_h is known, as in practice, but also the exact solution u. This allows the computation of exact nodal error numbers and error norms, to provide an objective assessment of the influence of mesh smoothing on solution efficiency and solution accuracy.

In the following, mesh smoothing, as applied in Step 2 of the finite element simulation process, is assessed by comparing smoothing iteration numbers, the smoothing time required, as well as the resulting mean ratio quality-based minimal and mean mesh quality numbers $q_{\min}$ and q_{mean}.

By using fine meshes and an associated large number of locally defined trial functions, the positive definite Ritz-Galerkin matrix G, also known as the stiffness matrix, becomes very large and sparse. The resulting system can efficiently be solved by the iterative conjugate gradient method (CG) [63] to determine the coefficient vector c. Therefore, the finite element solution efficiency of Step 4 can be assessed by considering the number of iterations required to achieve a prescribed solution accuracy and the resulting solution time. Furthermore, since numerical solution accuracy and efficiency are related to the condition number $\operatorname{cond} G$, as defined by (3.14), this number also serves as an indicator of the influence of mesh smoothing.

Finally, after Step 5 the approximate solution u_h can be compared with the prescribed exact solution u based on the nodal error numbers

$$e_{\max} := \max_{i \in I_\Omega} |u(p_i) - u_h(p_i)|, \tag{7.3}$$

$$e_{\mathrm{mean}} := \frac{1}{|I_\Omega|} \sum_{i \in I_\Omega} |u(p_i) - u_h(p_i)|, \tag{7.4}$$

where I_Ω denotes the index set of interior mesh nodes p_i and $|I_\Omega|$ the number of these nodes. Thus, $e_{\max}$ according to (7.3) represents the maximal nodal error and e_{mean} according to (7.4) the arithmetic mean of all nodal errors.

In Section 3.2.3 the approximate solution is derived as a weak solution of the variational formulation in appropriate Sobolev spaces. Therefore, the associated function-based error norms

$$e_{L^2} := \|u - u_h\|_{L^2(\Omega)} \quad \text{and} \quad e_{H^1} := \|u - u_h\|_{H^1(\Omega)} \tag{7.5}$$

are also of interest. Here, the L^2-error norm according to (3.3) represents the error based on the difference function $u - u_h$. Compared to this, the H^1-error according to (3.5) additionally incorporates differences in the first partial derivatives of $u - u_h$. These error numbers are well suited to demonstrating the solution accuracy for different types of trial functions defined over meshes with different element types.

7.4.2   Planar example

The first example considers the involute gear with 17 teeth depicted in Figure 7.35a as planar simulation domain Ω. The model has a diameter of about 19 units and is centered at the origin. The strongly varying exact solution

$$u(x,y) := 60 + 40 \sin \frac{(x+1)(y-1)}{2} \tag{7.6}$$

has been prescribed to better demonstrate the impact of mesh smoothing. In Figure 7.35b, the gear model is depicted together with a graph of the exact solution u. By applying the Laplace operator and interchanging signs, the right-hand side of the partial differential equation (7.1) can be derived as

$$f(x,y) := 10\big(x^2 + y^2 + 2(x - y + 1)\big) \sin \frac{(x+1)(y-1)}{2}.$$

The inhomogeneous boundary condition g is given by the restriction of u to the boundary $\partial\Omega$.

Two different polygonal meshes have been generated for Ω. One is a triangular mesh consisting of 14,346 elements sharing 7,660 nodes. 982 of these

(a) Model (b) Exact solution u

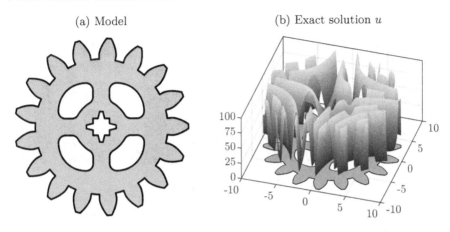

Figure 7.35: Involute gear model and exact solution u.

nodes are boundary nodes used for prescribing the inhomogeneous boundary condition by evaluating u at these locations. In consequence, nodal errors will be computed based on the remaining 6,678 interior nodes. The second mesh consists of 6,716 nodes and 6,229 quadrilateral elements. For better comparison, the quadrilateral mesh has the same boundary nodes as the triangular mesh, but differs in its interior nodes and node connectivity. Furthermore, the number of elements has been chosen such that the average element edge length is similar in both cases. Both meshes have been distorted after mesh generation by element validity-preserving random node movements, to demonstrate the influence of low mesh quality on finite element solution efficiency and accuracy.

Table 7.12: Triangular and quadrilateral gear mesh smoothing results.

Mesh	Method	Iter	Time (s)	q_{min}	q_{mean}
Tri	Initial	–	–	0.0000	0.3699
	Smart Laplace	18	0.23	0.0000	0.8790
	Global optimization	65	0.54	0.0000	0.4977
	GETMe	37 / 3,600	0.27	0.3369	0.8896
Quad	Initial	–	–	0.0004	0.4180
	Smart Laplace	27	0.15	0.0050	0.9394
	Global optimization	77	0.71	0.0045	0.5596
	GETMe	27 / 9,800	0.09	0.7902	0.9698

The initial mesh quality numbers as well as the smoothing results obtained by smart Laplacian smoothing, global optimization, and GETMe smoothing are given in Table 7.12 for both mesh types. Here, the same smoothing and termination parameters have been used as described in Section 7.1.1. As can be seen, both initial meshes are of very low quality due to mesh distortion. This also holds for the minimal element quality number q_{min} after applying

Triangular mesh

(a) Initial   (b) Smart Laplace   (c) Global opt   (d) GETMe

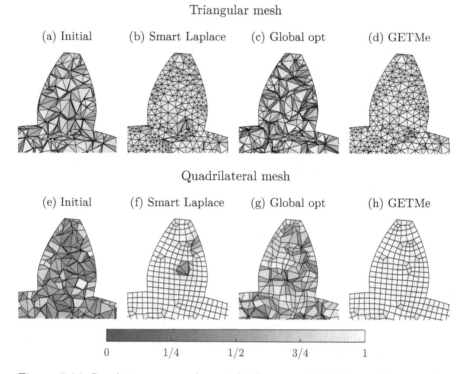

Quadrilateral mesh

(e) Initial   (f) Smart Laplace   (g) Global opt   (h) GETMe

0   1/4   1/2   3/4   1

Figure 7.36: Involute gear meshes with elements shaded according to their mean ratio quality number.

smart Laplacian smoothing, and it is caused by the clusters of low quality elements visible in the resulting mesh parts depicted in the Figures 7.36b and 7.36f. These remain, since a simple node resetting technique in the case of invalid element generation is applied to guarantee mesh validity. Applying adaptive local relaxation in node movement in the case of invalid element generation might prevent these effects, but would come at the expense of a higher implementational and computational effort. In contrast to the minimal element quality numbers, smart Laplacian smoothing increases the mean mesh quality numbers q_{mean} significantly in both cases.

Although global optimization conducts a comparably large number of iterations, the resulting mesh quality is low with respect to both quality numbers, as can be seen in Figures 7.36c and 7.36g and in the smoothing results given in Table 7.12. Here, the original termination criterion of the feasible Newton-based iteration of Mesquite has been applied, resulting in a preliminary termination with respect to the achieved quality improvements. Nevertheless, the total global optimization smoothing time is at least twice the time required by the other smoothing methods. In the case of the quadrilateral

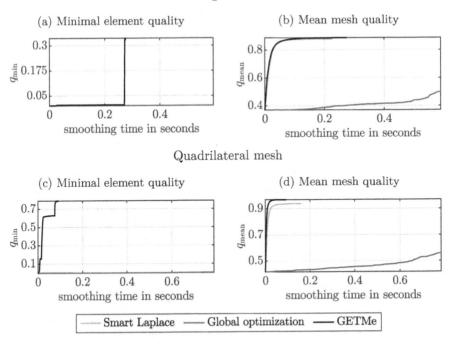

Figure 7.37: Involute gear mesh quality numbers with respect to smoothing time.

mesh, global optimization even takes almost eight times longer if compared to GETMe smoothing.

Finally, due to its combined approach of a simultaneous followed by a sequential substep, GETMe smoothing yields convincing results with respect to both quality numbers. Here, the better results of the quadrilateral mesh are due to a more suitable topological configuration if compared to the triangular mesh. That is, in the case of the quadrilateral mesh the number of elements attached to one inner node equals more often the ideal number of four if compared to the triangular mesh with an ideal number of six attached elements per inner node. This is also reflected by the lower iteration numbers for the simultaneous substep. The resulting high mesh quality is also visible from the mesh parts depicted in the Figures 7.36d and 7.36h.

Figure 7.37 depicts the resulting mesh quality numbers with respect to the smoothing time in seconds. Results for the triangular mesh are depicted in the upper row. It is visible from Figure 7.37a that the minimal element quality number is mainly improved by the sequential GETMe substep conducted in the second phase of GETMe smoothing. In contrast, the mean mesh quality number is improved by the simultaneous substep applied first, as is shown in Figure 7.37b.

Here, smart Laplacian smoothing and GETMe smoothing show the characteristic steep ascent of q_{mean} within the first few smoothing steps. In contrast, global optimization only leads to minor quality improvements during the first 0.5 smoothing seconds. As described before, the global optimization-based approach is terminated by its own termination criterion based on the inverse of the mean ratio quality number.

As is shown in Figure 7.37c, the minimal element quality number of the quadrilateral mesh is already significantly improved to a value greater than 0.6 during the simultaneous GETMe substep. Applying the sequential substep afterwards leads to another significant increase towards its final value of 0.7902. In contrast, as in the case of the triangular mesh, smart Laplacian smoothing and global optimization are not able to improve the minimal quality number considerably. In the case of the mean mesh quality number depicted in Figure 7.37d, smart Laplacian smoothing and GETMe smoothing show an even better quality improvement within the first few steps if compared to the triangular mesh case. For example, GETMe achieves a mean mesh quality number of 0.94 after applying only five iterations taking 0.0148 seconds in total.

The finite element solution u_h, approximating the prescribed analytic solution u, the nodal errors e_{max}, e_{mean}, and the Sobolev errors e_{L^2}, e_{H^1} have been computed by a solver based on the GetFEM++ library [59]. This package also provides the conjugate gradient solver used for solving the Galerkin system using a termination tolerance of 10^{-8} to determine the coefficient vector c of the trial functions modeling the approximate solution u_h. Preconditioning has not been applied to better demonstrate the impact of mesh quality on solution efficiency. Furthermore, the 2-norm condition numbers $\operatorname{cond} G$ of the resulting Galerkin matrices G have been computed by numerical determination of the smallest and largest eigenvalue $\lambda_{min}(G)$ and $\lambda_{max}(G)$, which yield $\operatorname{cond} G = \lambda_{max}(G)/\lambda_{min}(G)$.

Dirichlet-type boundary conditions according to (7.2) have been applied, where $g := u|_{\partial\Omega}$ was derived as the restriction of the prescribed analytic solution u to the boundary $\partial\Omega$. That is, boundary conditions were determined by evaluating (7.6) for all boundary nodes p_i. Since these have been kept fixed during smoothing, boundary conditions are identical for all meshes.

The finite element analysis has been accomplished for the piecewise linear ($d = 1$) elements tri3 and quad4 as well as for the piecewise quadratic ($d = 2$) Lagrangian elements tri6 and quad9, respectively. In doing so, the resulting number of degrees of freedom for the triangular and quadrilateral mesh amount to 7,660 and 6,716, respectively, for the linear case. For the quadratic basis, using six nodes triangular Lagrangian elements and nine nodes quadrilateral Lagrangian elements, the corresponding number of degrees of freedom increased to 29,670 and 25,894, respectively. The number of degrees of freedom matches the number of rows and columns of the associated Galerkin matrix G. As described in Section 3.2.5 exemplarily for linear elements, G is a sparse matrix, i.e., most of its entries are zeros. This is visible in Table 7.13, which provides the size of the Galerkin matrix G, its maximal number of nonzero entries per line, denoted as bandwidth, and its percentage of nonzero entries for the basis functions under consideration. The given data only depends on the type of the basis functions and the mesh topology. Thus, it is independent of mesh quality and the smoothing methods under consideration.

Table 7.13: Involute gear Galerkin matrix density information.

Mesh	Basis	Size	Bandwidth	Density in %
Tri	$d = 1$	7,660	13	0.0765
	$d = 2$	29,670	37	0.0347
Quad	$d = 1$	6,716	13	0.1098
	$d = 2$	25,894	37	0.0544

In contrast to the sparsity pattern of G, the entries of the Galerkin matrix depend on smoothing results, since node locations have an influence on the supports and shapes of the trial functions and, with this, on the integration domains of the entries of the Galerkin matrix derived according to (3.12). This is reflected by the condition number of G for the given initial meshes as well as their smoothed counterparts using smart Laplacian smoothing, global optimization, and GETMe smoothing. Results are given in Table 7.14 together with the conjugate gradient solver iteration numbers and the time required to solve the linear system $Gc = \hat{f}$.

Table 7.14: Involute gear CG solver results.

Mesh	Basis	Method	cond G	Iter	Time (s)
Tri	$d = 1$	Initial	1.09e+08	10,148	1.29
		Smart Laplace	4.28e+06	1,230	0.15
		Global opt	9.92e+05	2,069	0.27
		GETMe	4.60e+02	160	0.02
	$d = 2$	Initial	1.50e+09	43,449	68.25
		Smart Laplace	3.45e+07	8,370	14.29
		Global opt	1.26e+07	8,310	14.14
		GETMe	3.06e+03	430	0.60
Quad	$d = 1$	Initial	1.64e+05	1,452	0.17
		Smart Laplace	1.33e+04	390	0.05
		Global opt	1.48e+04	598	0.07
		GETMe	1.35e+02	93	0.01
	$d = 2$	Initial	2.74e+06	5,571	7.45
		Smart Laplace	2.19e+05	1,573	2.62
		Global opt	2.32e+05	2,169	2.18
		GETMe	1.14e+03	259	0.36

The impact of mesh smoothing on solution efficiency is particularly visible for the triangular mesh and GETMe smoothing. For linear basis functions, the condition number of G for the initial mesh is about 236,956 times larger than the condition number of G after applying GETMe smoothing. That is, GETMe smoothing leads to a significantly better conditioned linear system of equations, which can be solved by 160 CG iterations taking 0.02 seconds in total. Compared to this, solving the Galerkin system for the initial mesh takes about 64 times longer, since the number of CG iterations required is increased by almost the same factor.

For quadratic basis functions, the condition number of G for the initial mesh is 490,196 times larger if compared to cond G for the mesh after applying GETMe smoothing. Solving the linear system is speeded up by a factor of about 114. That is, by applying GETMe smoothing taking 0.09 seconds, the solution time is reduced by 67.65 seconds. Hence, applying GETMe smoothing leads to a significantly reduced total finite element solution time. A similar positive effect of GETMe smoothing on solution efficiency is also visible for the quadrilateral mesh, for both the linear as well as the quadratic basis functions. However, if compared to the results for the triangular mesh, the reduction factors for the condition number and solution times are smaller. This is due to a better topological configuration of the initial quadrilateral mesh.

By comparing the mesh quality numbers given in Table 7.12 and the condition numbers given in Table 7.14, it can be seen that smart Laplacian smoothing, although reaching a level similar to that of GETMe smoothing with respect to the mean mesh quality, is not able to result in similar condition numbers and solution times. For example, in the case of the triangular mesh and quadratic basis functions, the condition number for the mesh obtained by smart Laplacian smoothing is still 11,275 times larger if compared to those of GETMe smoothing. Furthermore, solving the Galerkin system takes about 24 times longer if compared to GETMe smoothing. Similar factors hold for the mesh obtained by applying global optimization.

The impact of mesh smoothing on Galerkin system solution efficiency becomes even more apparent if visualized with respect to the mesh smoothing time as depicted in Figure 7.38. In practice, the finite element solution process is only applied once for the final mesh after smoothing. However, to demonstrate the impact of mesh smoothing, for these graphs the finite element solution process has been applied after each smoothing iteration. This provides a deeper insight into the effects of mesh smoothing on solution efficiency. The left column shows the graphs for the smoothing methods under consideration for linear basis functions. Results for the quadratic basis functions are depicted on the right.

Results for the condition number cond G are given for the triangular mesh in the upper graphs. As can be seen, the first few smoothing iterations already decrease cond G by several orders of magnitude. However, this does not imply that the number of unconditioned CG iterations are decreased to the same extent. For example, applying the global optimization-based approach even leads to an increase of CG iteration numbers for the meshes obtained by the

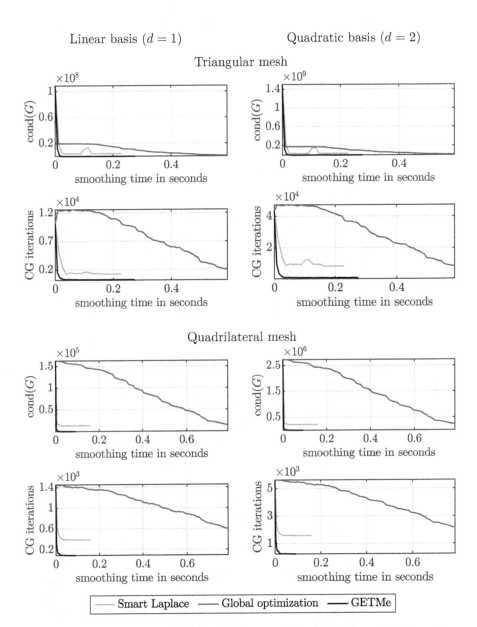

Figure 7.38: Involute gear Galerkin matrix condition numbers and CG iterations with respect to smoothing time.

first few smoothing iterations. After that, the CG iteration numbers stagnate on a very high level before they are significantly decreased. In contrast, the first few iterations of smart Laplacian smoothing and GETMe smoothing lead to a remarkable reduction of CG iterations. This correlates with a significant mesh quality improvement by the first few smoothing steps, as depicted in Figure 7.37.

For the quadrilateral mesh, global optimization-based smoothing is not able to reduce the Galerkin matrix condition numbers to the same extent within the first smoothing iterations as in the case of the triangular mesh. In contrast, with respect to solution efficiency, a few iterations of smart Laplacian and GETMe smoothing suffice to reduce solution times drastically. This is visible from the graphs depicted in the lower row of Figure 7.38. For example, starting from a total number of 5,571 CG iterations for the initial mesh, the GETMe simultaneous smoothing substep reduces the number of CG iterations to 1,312, 753, and 507 by the first, second, and third smoothing iterations.

The positive effect of GETMe smoothing not only holds for solution efficiency, but also for solution accuracy, as is depicted in Table 7.15. Here, the finite element solution errors are given for the triangular and quadrilateral meshes using linear and quadratic basis functions. The table lists the maximal and average nodal errors $e_{\max}$ and e_{mean} according to (7.3) and (7.4), as well as the Sobolev norm errors e_{L^2} and e_{H^1} according to (7.5).

Table 7.15: Involute gear finite element solution errors.

Mesh	Basis	Method	$e_{\max}$	e_{mean}	e_{L^2}	e_{H^1}
Tri	$d = 1$	Initial	31.2934	3.0878	54.0285	1,097.9648
		Smart Laplace	9.4938	0.8919	15.0708	101.5688
		Global opt	29.1443	2.4037	44.2167	468.7394
		GETMe	6.0790	0.8126	13.1683	79.1571
	$d = 2$	Initial	9.0099	0.1738	3.9982	179.9070
		Smart Laplace	0.9056	0.0163	0.4125	14.1967
		Global opt	3.4123	0.1161	2.6804	99.6673
		GETMe	0.5449	0.0132	0.3163	7.9146
Quad	$d = 1$	Initial	19.3538	2.3280	38.5825	309.5934
		Smart Laplace	9.7240	0.6560	11.4813	81.8776
		Global opt	23.6306	1.8151	32.6415	203.7577
		GETMe	4.3576	0.5813	9.6270	43.6017
	$d = 2$	Initial	2.3459	0.1025	2.0842	85.1985
		Smart Laplace	2.2336	0.0128	0.6341	17.0958
		Global opt	2.3257	0.0827	1.6912	48.1445
		GETMe	0.0751	0.0040	0.0777	1.3558

For example, in the case of the quadrilateral mesh using quadratic basis functions, the error numbers $e_{\max}$, e_{mean}, e_{L^2}, and e_{H^1} for the initial mesh are 31, 26, 27, and 63 times larger, respectively, if compared to the results for the mesh smoothed by GETMe. That is, smoothing the mesh leads to finite element solutions with significantly increased accuracy.

Table 7.15 also shows that for linear basis functions and the quadrilateral mesh, applying global optimization can even lead to a larger maximal nodal error if compared to the result for the initial mesh. Here, in practice, the preliminary termination of the feasible Newton approach of Mesquite should be circumvented by applying an alternative termination criterion. However, this comes at the expense of significantly larger iteration numbers, as has been demonstrated in [162]. By resulting in high quality meshes within short smoothing times, GETMe combines both benefits and thus results in a significantly improved solution efficiency and accuracy.

This is particularly visible in the graphs showing the finite element solution errors with respect to smoothing time for the triangular mesh depicted in Figure 7.39. In all graphs, GETMe smoothing leads to a significant reduction of the finite element solution errors within the first few smoothing steps of the simultaneous substep. Furthermore, the sequential substep has no significant impact on solution accuracy. That is, for the given example it would have been sufficient to apply about one third of the final number of GETMe simultaneous substeps to yield results comparable to those of applying the entire GETMe smoothing approach. Hence, in practice, smoothing time can be reduced even further.

Except for the H^1 error, applying smart Laplacian smoothing shows a similar behavior. As can be seen in the last row of Figure 7.39, there is a significant increase of the H^1 error in the middle of the smoothing phase. Comparing the H^1 error graphs with those of the L^2 error, which do not show such an increase, implies that these errors are caused by the partial derivatives of the finite element solution. Like the moderate mesh improvement effect of global optimization per iteration step, finite element solution error reduction is inferior. This holds in particular for the H^1 error, which stagnates or even increases during the second half of the smoothing process.

Figure 7.40 depicts the finite element solution errors with respect to smoothing time for the quadrilateral mesh. Again, results are given for linear as well as quadratic basis functions in the left and right column, respectively. As in the case of the triangular mesh, applying only a few GETMe simultaneous steps would already suffice to result in similar high quality results as applying the entire GETMe smoothing process. In contrast, global optimization leads to an increase of the maximal nodal error for both basis function types if compared to the initial mesh. Although the results of smart Laplacian smoothing do not show a temporary increase of the H^1 error as in the case of the triangular mesh, the resulting error numbers are larger in all cases if compared to the results obtained by GETMe smoothing. This holds in particular for the Sobolev norm-based error numbers.

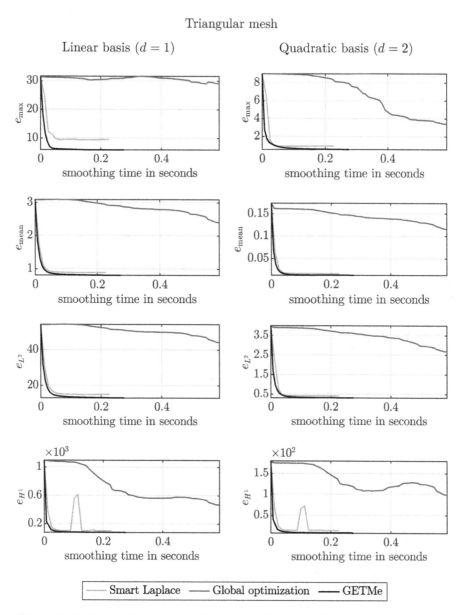

Figure 7.39: Finite element solution errors for triangular involute gear mesh with respect to smoothing time.

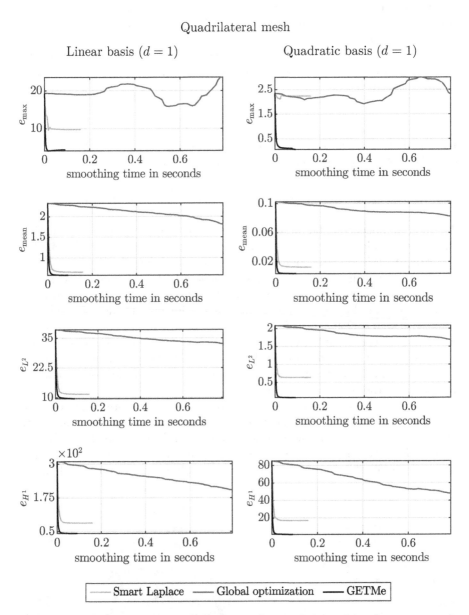

Figure 7.40: Finite element solution errors for quadrilateral involute gear mesh with respect to smoothing time.

7.4.3   Volumetric example

In the following, the impact of mesh smoothing on volumetric finite element solution efficiency and accuracy is analyzed using the example of the flange model depicted in Figure 7.41a. The height of the model amounts to 9 units and the diameter of the circular ground plate amounts to 16 units. As in the case of the involute gear model, an analytic solution of the Poisson problem (7.1) is prescribed by

$$u(x, y, z) := 35 \sin(2x + y) \cos(y - 3z) + 60. \tag{7.7}$$

By applying the Laplace operator and interchanging signs, the right-hand side of (7.1) used for the finite element simulation results in

$$f(x, y, z) := \frac{595}{2} \sin(2x + 2y - 3z) + \frac{455}{2} \sin(2x + 3z).$$

The non-homogeneous boundary conditions (7.2) are derived by evaluating the exact solution (7.7) at the boundary nodes. Slices of the model shaded according to this analytic solution are depicted in Figure 7.41b.

Two different initial meshes have been generated for the flange model. One is a tetrahedral mesh consisting of 103,230 nodes and 566,004 elements. Here, 18,961 nodes are located on the model surface for prescribing the boundary conditions, and 84,269 nodes are interior mesh nodes. The other is a hexahedral mesh with a comparable average element edge length consisting of 67,055 elements and 74,730 nodes, of which 15,154 are located on the model surface. That is, the number of nodes for prescribing boundary conditions is lower if compared to the tetrahedral mesh.

Due to the increased approximation power of the associated finite element basis functions, hexahedral elements are known to result in more accurate finite element solutions with a lower number of elements compared to tetrahedral meshes. However, depending on the geometrical complexity of the model, the generation of quality hexahedral meshes is usually far more complex than the

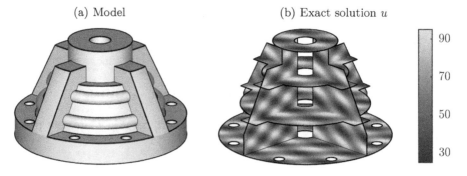

(a) Model   (b) Exact solution u

Figure 7.41: Flange model and slices shaded by the exact solution u.

generation of quality tetrahedral meshes. Since the flange model can easily be split into simple geometric primitives, it has been meshed by a sweeping approach. That is, each simple subdomain has been meshed by extruding a quadrilateral mesh along a sweeping path. Since the subdomains are connected by consistent quadrilateral interface meshes, a valid mesh is generated for the entire simulation domain. Compared to this, generating a tetrahedral volume mesh by Delaunay-based methods is far more flexible and efficient. Both meshes have been distorted by validity-preserving random node movements resulting in the mesh quality numbers given in Table 7.16.

Table 7.16: Tetrahedral and hexahedral flange mesh smoothing results.

Mesh	Method	Iter	Time (s)	q_{min}	q_{mean}
Tet	Initial	–	–	0.0001	0.4178
	Smart Laplace (C++)	19	23.37	0.0018	0.8745
	Global opt (C++)	33	18.01	0.0003	0.6634
	GETMe adaptive (C)	25 / 467	6.41	0.5967	0.8874
Hex	Initial	–	–	0.0606	0.4631
	Smart Laplace (C++)	28	13.82	0.0690	0.8777
	Global opt (C++)	49	17.25	0.1892	0.8403
	GETMe adaptive (C)	33 / 70	1.85	0.6339	0.9363

The initial meshes have been smoothed by smart Laplacian smoothing and global optimization using the same algorithms implemented in C++ as described in the previous sections. In addition, the initial meshes have also been smoothed by the GETMe adaptive approach as described in Section 6.2.2. To demonstrate not only the smoothing, but also the computational efficiency of this method, a more performance-focused implementation based on the C programming language [80] was used. The GETMe adaptive mean quality cycle as well as the smart Laplacian smoothing approach was terminated as soon as the mean mesh quality improvement of two consecutive iterations dropped below 10^{-4}. For the feasible Newton-based optimization method, its own gradient norm-based termination criterion was used. In all smoothing methods, boundary nodes have been kept fixed during smoothing.

The termination limits for the number of iterations without improvement for the mean and min cycle of GETMe adaptive have been set to 2 and 5, respectively, and the node relaxation tables $(\varrho_1, \ldots, \varrho_k)$ to $(1, 1/4, 1/16, 0)$ and $(1/2, 1/4, 1/10, 1/100, 0)$, respectively. During the min cycle, 1% of the worst-quality elements and their direct neighbors have been selected for improvement. The opposite face normals-based tetrahedral transformation and the dual element hexahedral element transformation described in Section 5.2 have been used, with the transformation parameter $\sigma = 3/2$ being applied in both cases. Since the global optimization-based approach is not parallelized, non-parallel versions of all smoothing methods have been used in both cases. Hence, in practice significantly shorter smoothing times than those given in Table 7.16

Tetrahedral mesh

(a) Initial   (b) Smart Laplace   (c) Global opt   (d) GETMe adapt

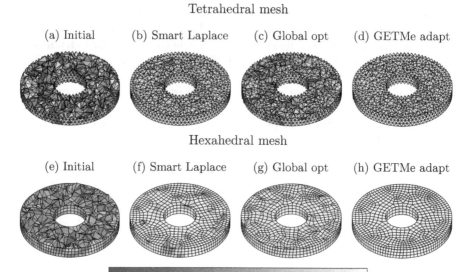

Hexahedral mesh

(e) Initial   (f) Smart Laplace   (g) Global opt   (h) GETMe adapt

0   1/4   1/2   3/4   1

Figure 7.42: Flange mesh cross sections with elements shaded according to their mean ratio quality number.

could be achieved for the test system described in the beginning of this chapter, by using parallelized algorithms.

As can be seen in the smoothing results table, the initial meshes are of low quality with respect to both minimal and mean element quality. This is also reflected by the mesh cross sections of the upper cylinder of the model depicted on the left side of Figure 7.42. Here, a large number of low quality elements shaded dark gray are visible in the tetrahedral mesh. In contrast, the element qualities of the initial hexahedral mesh are slightly better since preserving mesh validity is more restrictive in the case of hexahedral meshes. As in previous examples, smart Laplacian smoothing improves the mean mesh quality significantly but fails in improving the minimal element quality considerably. This can be seen by the clusters of low quality elements in both mesh cross sections depicted in Figure 7.42b and Figure 7.42f.

Although the iteration numbers of the global optimization-based approach are the largest, results are not satisfactory due to the preliminary termination by the inverse mean ratio-based quality criterion. Further smoothing would result in significantly higher quality numbers but comes at the expense of larger computation times [162]. Hence, the cross sections depicted in Figure 7.42c and Figure 7.42g have the character of an intermediate result.

Best results are obtained by the GETMe adaptive approach performing 25 iterations in the mean cycle and 467 iterations within 66 min cycles in the case of the tetrahedral mesh. For the hexahedral mesh, GETMe adaptive

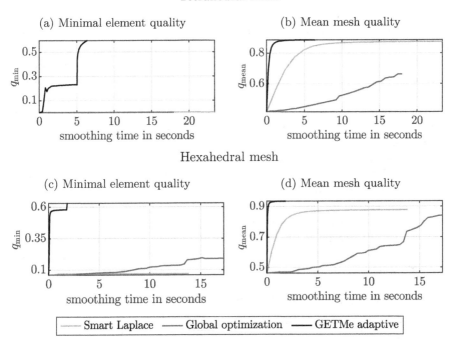

Figure 7.43: Flange mesh quality numbers with respect to smoothing time.

conducts 33 iterations and 70 iterations within one mean and 5 min cycles, respectively. The high quality is also reflected by the elements shaded light gray in the mesh cross sections depicted in Figure 7.42d and Figure 7.42h. By applying a GETMe adaptive version parallelized by OpenMP [103], the given runtimes are reduced by factor three on the given test system with four CPUs. That is, the hexahedral mesh is fully smoothed within less than one second. However, even in its non-parallel version, GETMe adaptive is more than nine times faster if compared with the global optimization-based approach.

Mesh quality numbers with respect to smoothing time are depicted in Figure 7.43. The left column shows the minimal element quality numbers for both mesh types. In the case of GETMe adaptive smoothing, the mean cycle increases the minimal element quality in both cases as is visible by the first quality plateau. After that, the minimal element quality is further increased by the min cycles applied afterwards. As can be seen, quality improvements by the mean cycle are better in the case of the hexahedral mesh if compared to the tetrahedral mesh. This is an observation made in several examples and caused by the more regular topological configurations generated by hexahedral mesh generators if compared to tetrahedral mesh generators. In contrast, for the tetrahedral mesh, both other methods fail to improve the minimal element quality number considerably. However, in the case of the hexahedral mesh, global optimization results in a moderate minimal element quality of $q_{min} = 0.1892$.

Better results are achieved for the mean mesh quality numbers as is depicted in Figure 7.43b and Figure 7.43d. Both smart Laplacian smoothing and GETMe adaptive smoothing show the characteristic steep ascent of q_{mean} within the first few smoothing steps. Here, using a performance-oriented implementation of smart Laplacian smoothing would result in a similar steep ascent as in the case of the GETMe smoothing. In contrast, mean mesh quality improvements of the feasible Newton approach of Mesquite are more moderate for almost all iterations.

As in the case of the involute gear model described in the previous section, the finite element solutions for the two different meshes of the flange model have been computed by the same GetFEM++-based solver implemented in C++. This is true for both linear basis functions ($d = 1$) using the elements tet4 and hex8, and for quadratic basis functions ($d = 2$) using the elements tet10 and hex27, respectively. The resulting degrees of freedom representing the matrix sizes, the bandwidths, and the density information of the associated Galerkin matrices are given in Table 7.17.

Table 7.17: Flange Galerkin matrix density information.

Mesh	Basis	Size	Bandwidth	Density in %
Tetrahedra	$d = 1$	103,230	23	0.0114
	$d = 2$	791,449	105	0.0031
Hexahedra	$d = 1$	74,730	33	0.0268
	$d = 2$	566,968	155	0.0094

As can be seen, compared to the tetrahedral case, the number of degrees of freedom of the Galerkin matrices for the hexahedral mesh are lower, due to the lower number of nodes and elements. However, the bandwidth is increased considerably. In both cases, the percentage of non-zero values is low. Again, the resulting Galerkin systems have been solved using the non-preconditioned CG solver also provided by the GetFEM++ package. In addition, the condition number $\text{cond}\,G$ has been computed for each Galerkin system. Results for the initial meshes as well as their smoothed counterparts are given in Table 7.18.

Comparable to the case of the triangular mesh in the involute gear example, the impact of mesh smoothing on the condition number $\text{cond}\,G$, and thus on the number of CG iterations, is particularly visible for the tetrahedral mesh. For example, applying GETMe adaptive smoothing reduces $\text{cond}\,G$ by five orders of magnitude for the linear as well as the quadratic basis. In consequence, the number of iterations required to solve the Galerkin system using a termination tolerance of 10^{-8} is reduced drastically. To be precise, the number of iterations decreases by the factors 54 and 128 in the linear and quadratic case, respectively. The associated solver runtimes are reduced by the factors 61 and 180, respectively.

Table 7.18: Flange CG solver results.

Mesh	Basis	Method	cond G	Iter	Time (s)
Tet	$d = 1$	Initial	2.34e+07	7,519	67.58
		Smart Laplace	1.30e+05	1,267	15.02
		Global opt	2.01e+06	860	10.42
		GETMe adaptive	3.60e+02	139	1.10
	$d = 2$	Initial	4.90e+08	42,312	5,067.10
		Smart Laplace	2.12e+06	6,021	570.65
		Global opt	3.87e+07	4,937	637.20
		GETMe adaptive	2.33e+03	331	28.11
Hex	$d = 1$	Initial	1.51e+04	508	3.07
		Smart Laplace	1.94e+03	228	1.36
		Global opt	5.07e+02	161	0.93
		GETMe adaptive	3.66e+02	141	0.82
	$d = 2$	Initial	7.35e+05	3,810	385.03
		Smart Laplace	2.16e+05	1,195	107.99
		Global opt	6.20e+03	531	42.09
		GETMe adaptive	3.59e+03	406	32.16

Larger condition numbers do not necessarily lead to larger CG iteration numbers. This can be seen for the tetrahedral mesh and the results obtained by smart Laplacian smoothing and global optimization. In both cases, the global optimization-based condition number is larger by one order of magnitude if compared to that obtained by applying smart Laplacian smoothing. However, the condition number gives an indication of solution efficiency, as is visible by the results for the initial and the smoothed meshes.

Due to its generation scheme based on sweeping, the hexahedral mesh has a more favorable topological configuration if compared to the tetrahedral mesh. This is also reflected by the associated condition numbers and solver results for the initial mesh given in the lower half of Table 7.18. In both cases, the iteration numbers are much lower if compared to those of the tetrahedral mesh. Nevertheless, GETMe adaptive smoothing reduces the initial mesh-related condition numbers by two orders of magnitude and with this the CG iteration numbers by factors 3.6 and 9.4, respectively, in the linear and quadratic cases. The CG solver time is reduced by factors 3.7 and 12.0, respectively.

To give a more detailed picture of the impact of mesh smoothing on finite element solution efficiency, the condition number and CG solver information has not only been computed for the final meshes obtained by all smoothing methods, but also for each intermediate mesh obtained by a single smoothing iteration. In addition, for each iteration mesh, a full finite element analysis has been conducted. The resulting condition number and number of CG iterations are depicted in Figure 7.44 for all smoothing methods with respect to the mesh

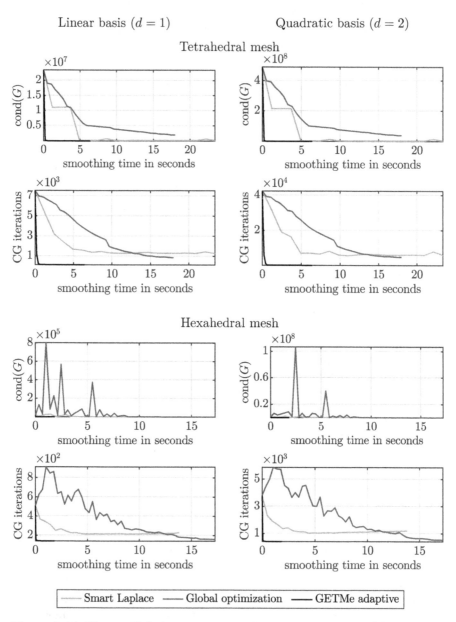

Figure 7.44: Flange Galerkin matrix condition numbers and CG iterations with respect to smoothing time.

smoothing time. This gives an indication of finite element solution efficiency, after a preliminary smoothing termination at a given time.

Results for linear basis functions are depicted on the left side, those for quadratic basis functions on the right. As can be seen, the first few GETMe adaptive iterations already have a remarkable impact on solution efficiency, as is visible by the steep descent in the graphs depicted in Figure 7.44. For example, the condition number 4.90e+08 associated with the initial tetrahedral mesh using piecewise quadratic basis functions is reduced by the first two iterations of GETMe adaptive smoothing to 3.90e+05. With this, the initial CG iteration number is reduced from 42,312 to 1,297. In the same case, two iterations of smart Laplacian smoothing result in the condition number 2.13e+08 and 19,081 CG iterations. Two iterations of global optimization result in the condition number 3.92e+08 and 38,804 CG iterations.

Results for the hexahedral mesh are depicted in the lower part of the same figure. In the case of the global optimization-based approach, condition numbers and CG iteration numbers are even increased during the first phase of mesh smoothing, if compared to the initial mesh results. For example, in the case of linear basis functions the maximal CG iteration number is 912, which is 1.8 times the initial value. In the case of quadratic basis functions, the maximal condition number is 1.07e+08, which is three orders of magnitude larger if compared to the initial value. Thus, for the given example, a preliminary termination of global optimization comes at the risk of an even lower solution efficiency as in the case of the initial mesh.

Table 7.19: Flange finite element solution errors.

Mesh	Basis	Method	e_{max}	e_{mean}	e_{L^2}	e_{H^1}
Tet	$d = 1$	Initial	26.5052	3.7173	129.4961	2,332.6902
		Smart Laplace	14.2884	0.8750	31.4086	271.7764
		Global opt	11.9840	1.6789	59.8793	418.5072
		GETMe adapt	4.5698	0.7996	27.8099	148.5165
	$d = 2$	Initial	4.3067	0.2119	8.8008	554.4774
		Smart Laplace	3.3903	0.0196	1.0404	52.3666
		Global opt	3.2543	0.0649	2.7483	68.9090
		GETMe adapt	0.2967	0.0142	0.5216	13.2547
Hex	$d = 1$	Initial	16.1917	2.0114	66.1737	466.4109
		Smart Laplace	8.9167	0.6372	23.2996	126.0352
		Global opt	6.3404	0.7067	25.2152	119.9322
		GETMe adapt	2.2328	0.4478	15.1903	66.8219
	$d = 2$	Initial	3.7028	0.1058	4.8893	105.2790
		Smart Laplace	2.4930	0.0133	0.9654	18.6139
		Global opt	2.1401	0.0142	0.7921	13.8879
		GETMe adapt	0.6024	0.0042	0.2390	3.0018

Table 7.19 gives the maximal and average nodal errors e_{max} and e_{mean}

according to (7.3) and (7.4), as well as the Sobolev norm errors e_{L^2} and e_{H^1} according to (7.5) for the initial meshes as well as their smoothed counterparts. As in the case of the involute gear model, the use of quadratic basis functions instead of linear basis functions reduces all error numbers due to their increased approximation power. Similarly, results for hexahedral meshes are better if compared to the results obtained for tetrahedral meshes. However, this only holds for quality meshes, since, for example, errors for the GETMe adaptive smoothed tetrahedral mesh are smaller than those of the initial hexahedral mesh.

The lower mesh quality numbers of the tetrahedral global optimization-based mesh are reflected by higher solution errors. For example, global optimization reduces the e_{H^1} errors for linear and quadratic tetrahedral elements by factors 5.6 and 8.0 if compared to the initial mesh. In contrast, the reduction factors achieved by GETMe adaptive smoothing amount to 15.7 and 41.8, respectively. Similarly, smart Laplacian smoothing reduces the maximal nodal error for quadratic basis functions by the factors 1.3 and 1.5, respectively, in the case of tetrahedral and hexahedral elements. The reduction factors achieved by GETMe adaptive smoothing for the same cases amount to 14.5 and 6.1, respectively. That is, GETMe adaptive smoothing leads to convincing results with respect to all error numbers under consideration.

Similar to the case of the condition and CG iteration numbers, significant finite element solution error number reductions are achieved within the first few GETMe adaptive iterations. For the tetrahedral mesh, this is shown in Figure 7.45, depicting these error numbers with respect to smoothing time. Again, these graphs have been generated by conducting an entire finite element solution process for each iteration mesh of each smoothing method. In the case of quadratic basis functions, the maximal nodal error is slightly increased during the first third of global optimization-based smoothing, if compared to the initial nodal error.

Since all mesh elements are smoothed during the first stage of GETMe adaptive smoothing, this phase is similar to the GETMe simultaneous substep. Hence, both are particularly effective with respect to all approximation errors and basis types, as can be seen in all graphs. For the given examples, the second stage improving the minimal element quality has no further significant effect on finite element solution quality. Hence, from a practical point of view, applying only a few GETMe adaptive or GETMe simultaneous iterations suffice already in the given example.

In the case of the hexahedral mesh and quadratic basis functions, the second stage of GETMe adaptive smoothing slightly increases the maximal nodal error as is depicted in Figure 7.46. This is not the case for linear basis functions. Again, the steep descent of error numbers with respect to smoothing time achieved by GETMe adaptive smoothing is visible for all error and basis types. Using a C implementation of smart Laplacian smoothing would also lead to a similar steep descent. However, the final error numbers do not reach the high quality results obtained by GETMe adaptive smoothing.

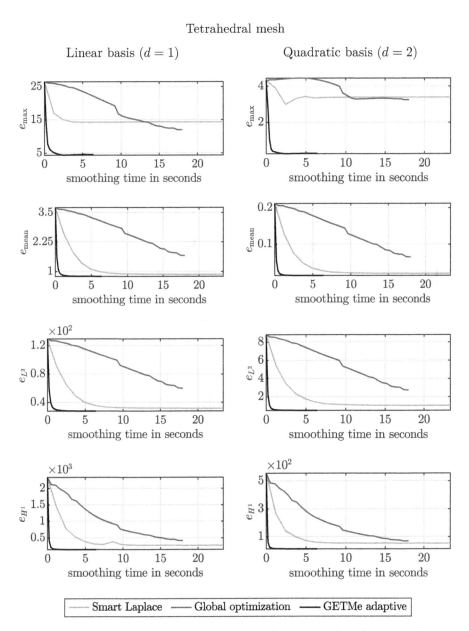

Figure 7.45: Finite element solution errors for tetrahedral flange mesh with respect to smoothing time.

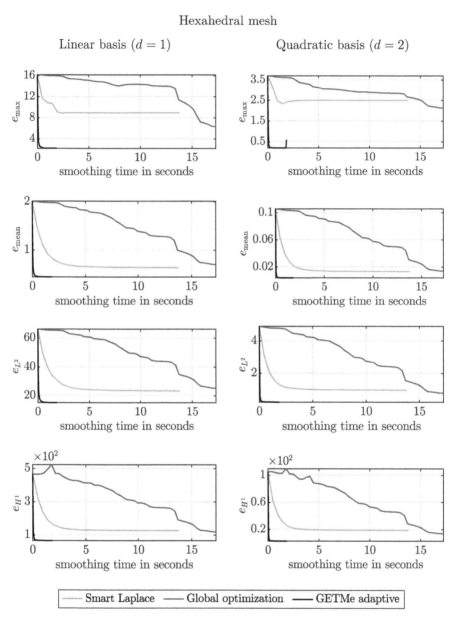

Figure 7.46: Finite element solution errors for hexahedral flange mesh with respect to smoothing time.

During global optimization-based hexahedral mesh smoothing, the condition numbers and CG iteration numbers have been increased remarkably, as was illustrated in Figure 7.44. However, this does not hold to the same extent for the approximation errors. Here, only the e_{H^1} errors are slightly increased within the first smoothing iterations. In practice, similar low error numbers as obtained by GETMe smoothing can be achieved by applying sufficiently many iterations of the global optimization-based approach. However, this comes at the expense of a further increased smoothing runtime.

All examples given in this section demonstrate that GETMe-based smoothing is highly efficient and applicable to all kinds of meshes. Furthermore, it is easy to implement and allows to improve both the mean mesh quality as well as the minimal element quality number by combining the simultaneous as well as the sequential approach. Furthermore, in its basic form it is even able to untangle invalid meshes with a suitable topology.

8

Extending the GETMe smoothing framework

The transformations and smoothing methods presented in the previous chapters are based on a geometric approach, and focus on computational efficiency, implementational simplicity, and controllability. Furthermore, the regularizing effect of polygon transformations has been proven. However, there is yet no proof of regularization for the polyhedral element transformations given in Section 5.2 and the aspects of GETMe smoothing. Therefore, this chapter provides definitions of alternative element transformations facilitating the mathematical analysis of element regularization and mesh improvement. Furthermore, an overview of recent research results regarding the theoretical background of geometric element transformation-based smoothing methods is given.

8.1 Alternative element transformations

8.1.1 Rotation-based transformations

A regularizing transformation for triangles based on imitating the rotational symmetry group action of the triangle is given in [145]. Let $p_0, p_1, p_2 \in \mathbb{R}^2$ denote the nodes of a counterclockwise-oriented triangle in the Euclidean plane. Its centroid serves as a rotation center. In the case of a regular triangle, rotating by angle $2\pi/3$ maps each node p_i on its successor node p_{i+1}, $i \in \{0, 1, 2\}$, where here and in the following all indices must be taken modulo n.

Instead of rotating nodes, the following transformation approach is based on transferring the distance of a node from the centroid towards its successor. In addition, it is also ensured by construction that the transformation preserves the centroid of the triangle. Combining both steps results in the following transformation.

Definition 8.1 (Rotation-based triangle transformation). *Let $c := (1/3)(p_0 + p_1 + p_2)$ denote the centroid of a counterclockwise-oriented triangle with nodes $p_0, p_1, p_2 \in \mathbb{R}^2$ and $r_i := \|p_{i-1} - c\|/\|p_i - c\|$, $i \in \{0, 1, 2\}$, the ratio of the distances of two consecutive triangle nodes from the centroid. The nodes of the transformed triangle are given by*

$$p_i' := \frac{2}{3}r_i(p_i - c) - \frac{1}{3}r_{i+1}(p_{i+1} - c) - \frac{1}{3}r_{i-1}(p_{i-1} - c) + c, \quad (8.1)$$

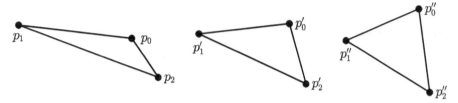

Figure 8.1: Example of iteratively applying the rotation-based triangle transformation and its regularizing effect.

where $\|\cdot\|$ denotes the Euclidean norm.

As can be seen by (8.1), the nodes after applying one step of the transformation can be represented as a weighted sum of centroid-to-node direction vectors scaled by distance relations. By deriving estimates for the minimal and maximal node distances it is shown in [145] that the ratios r_i tend to one, if the transformation is iteratively applied. This result also implies that iteratively applying the transformation given by Definition 8.1 results in regular triangles.

Figure 8.1 depicts on the left an initial triangle. It is transformed two times by the rotation-based triangle transformation, resulting in the triangles with nodes p_i' and p_i'' depicted in the middle and on the right, respectively. As can be seen, already two iterations lead to a significant improvement of shape towards a regular triangle.

The transformation given in Definition 8.1 is invariant under rotation, scaling, translation, and reflection [147]. Thus, it fulfills the minimal requirements for a suitable regularizing element transformation. In addition, it is applicable to triangles in $\mathbb{R}^3$ without modification. Here, the transformed triangle is located in the same plane as the initial triangle. Let Λ denote the set of similar triangles for a given equilateral triangle. By relating the transformation to a system of coupled damped oscillations, [147] gives a sketch of the proof that Λ is a global attractor of this transformation. Furthermore, for any planar triangle the sequence of iteratively transformed triangles converges uniformly at an exponential rate to an element of Λ.

The given triangle transformation can also be utilized as a building block for transforming other element types. To transform a quadrilateral element, it is subdivided into triangles along each of its two diagonals.

For the quadrilateral element with nodes $p_0, \ldots, p_3$ depicted on the left of Figure 8.2, this results in the four triangles with node indices $(0, 1, 2)$, $(2, 3, 0)$, $(0, 1, 3)$, and $(1, 2, 3)$, respectively. Each of these triangles is transformed according to Definition 8.1, resulting in the four triangles depicted in the middle of Figure 8.2. Here, the transformed triangles associated to the first diagonal are marked dark gray, those associated to the second diagonal are marked light gray. Each node of the initial quadrilateral is related to three nodes of the transformed subdivision triangles. For example, the transformed triangle nodes associated to p_3 are marked by discs in the middle of Figure 8.2. Each

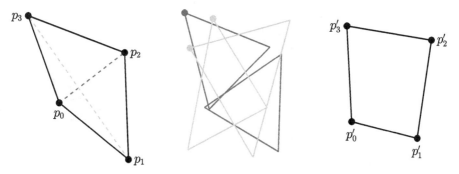

Figure 8.2: Transformation of a quadrilateral based on subdivision and applying the triangle transformation.

Figure 8.3: Transformation of a tetrahedral element by face transformations.

node p_i' of the transformed quadrilateral is derived as arithmetic mean of its associated transformed triangle nodes. For the given example, this results in the transformed quadrilateral element depicted on the right of Figure 8.2. Applying this transformation iteratively results in a sequence of quadrilaterals, which converges to a regular quadrilateral element.

Compared to the generalized regularizing polygon transformations given by Definition 5.1, there is no straightforward generalization to regularize polygons with an arbitrary number of nodes using the triangle-based transformation approach.

Tetrahedral elements are transformed by applying the triangle transformation to each face of the tetrahedron. Again, the nodes of the transformed tetrahedron are derived by taking the arithmetic mean of the associated three transformed triangle nodes. The transformation of a face might reverse its orientation. If this is the case, the original triangle is used instead of the transformed one [147]. The steps of this transformation are illustrated in Figure 8.3. The left side shows an initial tetrahedron. Each face is indicated by a slightly shrunken triangle with an individual shade. The resulting triangles, after applying the transformation according to Definition 8.1, are depicted in the middle, using the same shades as on the left side. Again, transformed nodes associated to p_3 are marked by discs. The resulting transformed tetrahedron

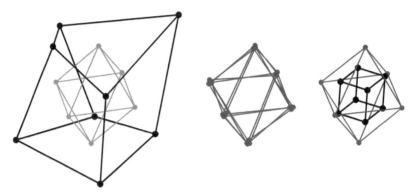

Figure 8.4: Transformation of a hexahedron by dual octahedron face transformations.

obtained by transformed face-triangle nodes averaging is shown on the right of Figure 8.3.

Based on this transformation, [146] gives a proof for the existence of a local attractor of the tetrahedron transformation, which coincides with the set of regular tetrahedra. It is founded on defining the space of tetrahedra under transformation as the homogeneous space $SL(3, \mathbb{R})/SO(3, \mathbb{R})$. Here, by representing a tetrahedron by the matrix of its spanning vectors, the set of tetrahedra is defined as the special linear group $SL(3, \mathbb{R})$. Furthermore, it is used that the transformation commutes with the rotational group $SO(3, \mathbb{R})$. The proof given in [146] incorporates techniques from the theory of compact group actions and the geometric structure of Lie groups and their Lie algebras together with results of the theory of dynamical systems.

Picking up the idea of the dual element-based volumetric element transformations described in Section 5.2.2, hexahedral elements can also be transformed by the given triangle transformation as is depicted in Figure 8.4. First, the nodes of the dual octahedron are derived as the face centroids of the hexahedron to be transformed. The left side of Figure 8.4 shows the initial hexahedron marked black and the resulting dual octahedron marked light gray. Each triangular face of the dual octahedron is transformed by the triangle transformation. The resulting eight transformed unconnected triangle faces are depicted in the middle of Figure 8.4. Each node of the transformed octahedron is derived by taking the arithmetic mean of the four associated transformed triangle nodes. The resulting transformed octahedron is marked dark gray on the right side of Figure 8.4. Finally, the transformed hexahedron marked black is determined as the dual element of the transformed octahedron, i.e., the transformed hexahedron nodes are the centroids of the transformed octahedron face nodes. As in the case of the element transformations described in Chapter 5, scaling techniques can be applied afterwards to preserve the size of the original element.

8.1.2 Geometric transformations with non-regular limits

In Section 5.1.3 an algebraic approach has been presented to derive customized polygon transformations leading to limit polygons with a prescribed not necessarily regular shape. For polygons with an arbitrary number of nodes this was done by deriving the transformation matrix from a suitable choice of eigenvalues and eigenvectors, where the eigenvector belonging to the dominant eigenvalue represents the target shape.

Following the spirit of the generalized regularizing polygon transformations described in Section 5.1.2, the alternative transformation scheme presented in [152] proposes a geometric approach also resulting in non-regular limit polygons with a given shape. Here, a polygon $z \in \mathbb{C}^n$ with $n \geq 3$ nodes z_k is transformed according to $z' := Mz$, where

$$
M := \begin{pmatrix}
1 - w_0 & w_0 & 0 & \cdots & & 0 \\
0 & 1 - w_1 & w_1 & \cdots & & 0 \\
\vdots & & \ddots & \ddots & & \vdots \\
0 & \cdots & 0 & 1 - w_{n-2} & w_{n-2} \\
w_{n-1} & 0 & \cdots & & 0 & 1 - w_{n-1}
\end{pmatrix}, \tag{8.2}
$$

with $z_k, w_k \in \mathbb{C}$ for $k \in \{0, \ldots, n-1\}$ using zero-based indices taken modulo n. As can be seen, M has the same sparsity pattern as the Matrix M'' defined by (5.3). However, it is not circulant, since for each transformed node $z'_k := (1 - w_k)z_k + w_k z_{k+1}$ an individual transformation parameter w_k is used. Due to $z'_k = z_k + w_k(z_{k+1} - z_k)$, this represents the construction by rotating a scaled version of the edge $e(z_k, z_{k+1})$ around z_k by an angle $\theta_k = \arg(w_k)$ using the scaling factor $\lambda_k = |w_k|$. It holds in particular that $\theta_k \in [0, 2\pi)$ and $\lambda_k \in \mathbb{R}$.

The matrix of the regularizing polygon transformation given in Section 5.1.2 is circulant, which plays a key role in the analysis of its limit behavior. Being circulant implies that M can be diagonalized by the discrete Fourier matrix independent of its construction parameters λ and θ. Here, the columns of the discrete Fourier matrix, called Fourier polygons, result in a full classification of the limit polygons obtained by iteratively applying this transformation. As in the case of the regularizing polygon transformation, limit polygons obtained by applying the transformation according to (8.2) can be determined by deriving the associated eigenvalues and eigenvectors of M. However, since M is non-circulant, these depend on the transformation parameters w_k.

Let I denote the $n \times n$ identity matrix. For any eigenvector-eigenvalue pair $v_k \in \mathbb{C}^n$, $\eta_k \in \mathbb{C}$ of M, i.e., $Mv_k = \eta_k v_k$, it holds that $(M - I)v_k = (\eta_k - 1)v_k$. That is, $\eta_k - 1$ is an eigenvalue of $M - I$ with eigenvector v_k. It is shown in [152] that the characteristic polynomial of $M - I$ is given by

$$
p(\eta) = \prod_{k=0}^{n-1}(\eta + w_k) - \prod_{k=0}^{n-1} w_k. \tag{8.3}
$$

Since the second product of (8.3) becomes zero if at least one of the parameters w_k is zero, this case is of interest.

Assuming without loss of generality that $w_0 := 0$ and $w_k \neq 0$ for $k > 0$, the characteristic polynomial of $M - I$ simplifies to

$$p(\eta) = \eta \prod_{k=1}^{n-1} (\eta + w_k),$$

which implies that the eigenvalues of M are given by $\eta_k = 1 - w_k$. As can be easily verified, the first eigenvector of M is given by $v_0 := (1, \ldots, 1)^t$. Let $v_{j,k}$ be the jth entry of the kth eigenvector. For $k > 0$, the kth eigenvector is computed as the solution of the linear system $(M - \eta_k I)v_k = (M + (w_k - 1)I)v_k = 0$. The jth equation of this linear system results to

$$(w_k - w_j)v_{j,k} + w_j v_{j+1,k} = 0. \tag{8.4}$$

For $j = 0$ it follows that $v_{0,k} = 0$, since $w_0 = 0$. The same holds for $j = k$, which implies $v_{k+1,k} = 0$. Using that $w_j \neq 0$ for $j > 0$, the expression (8.4) can be written as $v_{j+1,k} = ((w_j - w_k)/w_j)v_{j,k}$. Together with the normalization $v_{1,k} := 1$ and the other entries already derived, this implies

$$v_{j,k} = \begin{cases} 0 & \text{if } j = 0 \text{ or } j > k \\ \prod_{\ell=1}^{j-1} \dfrac{w_\ell - w_k}{w_\ell} & \text{otherwise,} \end{cases}$$

with $k > 0$ and the common convention that empty products result in one.

Let V be the eigenvectors matrix with entries $v_{j,k}$ and D denoting the diagonal matrix of all eigenvalues η_k. As in the case of the other linear polygon transformations, it follows that $M = VDV^{-1}$ if V is invertible. Hence, applying the transformation ℓ times to an initial polygon $z^{(0)}$ results in $z^{(\ell)} = VD^\ell V^{-1} z^{(0)}$, and the eigenvector belonging to the eigenvalue with the largest magnitude determines the limit shape.

In the case of the customized polygon transformations described in Section 5.1.3, the eigenvectors have been chosen as the target polygon and its variations based on specific node permutations. In contrast, the geometric transformation considered in this section is customized by an appropriate choice of the node-related transformation parameters w_k. Let $u \in \mathbb{C}^n$ denote the prescribed target polygon. It is shown in [152] for the specific choice

$$\tilde{w}_k := u_k / (u_{k+1} - u_k) \tag{8.5}$$

that u is an eigenvector of the associated transformation matrix $\tilde{M}$ according to (8.2) with eigenvalue two. However, it cannot be assumed in general that two is the largest eigenvalue of $\tilde{M}$ with the remaining eigenvalues being denoted as $1 + \eta_k$. Therefore, using a complex number $\lambda \in \mathbb{C}$ the weighted coefficients $w_k := \lambda \tilde{w}_k$ and the resulting transformation matrix M are considered.

For this choice, the eigenvalues of M result in $1, 1+\lambda, 1+\lambda\eta_2, \ldots, 1+\lambda\eta_{n-1}$ with $1 + \lambda$ representing the eigenvalue of the target polygon u. Furthermore, $\tilde{M}$ and M share the same eigenvectors. The dominance of the eigenvalue $1 + \lambda$ is obtained by a suitable choice of λ. To do so for a prescribed complex number η the set $\Lambda_\eta := \{\lambda \in \mathbb{C} : |1 + \lambda|^2 > |1 + \lambda\eta|^2\}$ of λ values for which $1 + \lambda$ is dominant is considered. Depending on the absolute value of η this set is characterized in [152] as follows, where $\omega_\eta := (1 - \overline{\eta})/(|\eta|^2 - 1)$:

$|\eta| < 1$: Λ_η is the exterior of the circle with radius $|\omega_\eta|$ and center ω_η.

$|\eta| = 1$: Λ_η is the half plane of λ with angle to $\overline{\eta} - 1$ below $\pi/2$.

$|\eta| > 1$: Λ_η is the interior of the circle with radius $|\omega_\eta|$ and center ω_η.

It follows readily that if the intersection of all Λ_{η_k} for the remaining eigenvalues η_k, $k \in \{2, \ldots, n - 1\}$ is non-empty, suitable transformations exist. In this case M can be computed by taking λ from this intersection and deriving the parameters w_k as defined before. For the case $n = 4$ a specific choice of λ is proposed in [152] based on the classification of $|\eta_2|$ and $|\eta_3|$ described before.

For comparison reasons, this computation scheme is now demonstrated with the example already considered in Section 5.1.3, resulting in the target rectangle u depicted in Figure 5.16. For $\alpha = \pi/3$, the target polygon is given by $u = (1, e^{i\alpha}, -1, e^{i(\pi+\alpha)})^t$ representing a rectangle with side length ratio $= 1/\sqrt{3}$. Evaluating (8.5) for these nodes u_k, the parameters of $\tilde{M}$ result in $\tilde{w}_0 = \tilde{w}_2 = e^{-2\pi i/3}$ and $\tilde{w}_1 = \tilde{w}_3 = e^{-5\pi i/6}/\sqrt{3}$. The eigenvalues of $\tilde{M}$ are given by $1, 2, 1 + 2i/\sqrt{3}, 2 + 2i/\sqrt{3}$ with the first belonging to the eigenvector $v_0 = (1, \ldots, 1)^t$ and the second belonging to $v_1 = u$. It follows that $\eta_2 = 2i/\sqrt{3}$ and $\eta_3 = 1 + 2i/\sqrt{3}$ with absolute values both being greater than one. Thus, valid values for λ are located in the intersection of the open disks with centers $\omega_2 = 3 + 2\sqrt{3}i$ and $\omega_3 = \sqrt{3}i/2$ and radii given by the absolute values of these center points. Choosing $\lambda = i$ and the associated parameters $w_k = \tilde{w}_k i$ results in a transformation matrix M, which yields the prescribed shape of u if applied iteratively to an arbitrary initial polygon.

This is illustrated with the example of the initial polygons $z^{(0)}$, depicted on the left of Figure 8.5. To make transformation results comparable, these are the same as in the customized transformation example depicted in Figure 5.17. In both cases, the polygons obtained by iteratively applying the transformation are shown from left to right. Here, the first and second node of each polygon is marked by a circle and square marker, respectively, to indicate the orientation of the polygons. All remaining nodes are marked by black discs.

The transformation towards the prescribed target shape u is clearly visible in Figure 8.5. It can also be seen that the transformation obtained in this section has a rotational effect due to the complex eigenvalue with largest magnitude. In a mesh smoothing context, such rotations can effectively be compensated by applying the target angle modifications described in Section 5.1.3.

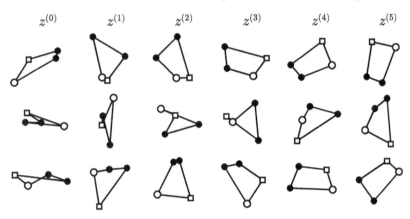

Figure 8.5: Examples of iteratively applying the geometric transformation with a non-regular target shape.

8.1.3 Gradient flow

The transformation of tetrahedral elements by opposite face normals, as described in Section 5.2.1, builds the foundation for alternative volume element transformations proposed in [149, 150]. In these publications, normals of faces bounded by arbitrarily oriented polygonal curves with nodes $p_1, \ldots, p_k \in \mathbb{R}^3$ are defined by the sum of their consecutive cross products

$$\nu(p_1, \ldots, p_k) := (p_1 \times p_2) + (p_2 \times p_3) + \cdots + (p_{k-1} \times p_k) + (p_k \times p_1).$$

If the polygonal curve is planar and convex, the Euclidean norm of the vector $\nu(p_1, \ldots, p_k)$ equals twice the area enclosed by the polygon [150].

Using this notation and the tetrahedral element node numbering scheme depicted in Figure 5.21, the column vector of all four opposite face normals can be written as

$$X_{E^{\text{tet}}} := \begin{pmatrix} \nu(p_4, p_3, p_2) \\ \nu(p_4, p_1, p_3) \\ \nu(p_4, p_2, p_1) \\ \nu(p_1, p_2, p_3) \end{pmatrix}. \tag{8.6}$$

The signed volume function of a tetrahedron $E^{\text{tet}} = (p_1, \ldots, p_4)$ is given by the sixth of the determinant of the matrix of its spanning vectors, i.e.,

$$\text{vol}(E^{\text{tet}}) := \frac{1}{6} \det(p_2 - p_1, p_3 - p_1, p_4 - p_1).$$

Using this, the gradient field of the tetrahedron volume can be written as $\nabla \text{vol}_{\text{tet}} = (1/6) X_{E^{\text{tet}}}$.

The main difference in defining the new regularizing tetrahedron transformation compared to the opposite face normals transformation given in Section 5.2.1 lays in the choice of normalization to ensure scale invariance of the transformation. Whereas for the opposite face normals transformation scale invariance has been obtained by normalizing each of the face normals individually, [149] proposes to normalize the vector $X_{E^{tet}}$ given by (8.6) based on the normalization function

$$\Psi(X) := \begin{cases} \frac{1}{\sqrt{\|X\|}}X & \text{if } \|X\| \neq 0 \\ 0 & \text{otherwise.} \end{cases}$$

Let $x := (p_1^t, \ldots, p_4^t)^t \in \mathbb{R}^{12}$ denote the column vector consisting of all four tetrahedron nodes and $\sigma > 0$ a scaling factor. The element nodes' coordinates vector $x' \in \mathbb{R}^{12}$ after applying one step of the new regularizing element transformation is given by

$$x' := x + \sigma\Psi\left(X_{E^{type}}\right). \tag{8.7}$$

This transformation scheme also applies to hexahedra, prisms, and pyramids, which only differ in the definition of the element type-dependent choice of the faces' normals vector $X_{E^{type}}$. These are derived from tessellations of these elements into tetrahedra sharing the same nodes as the original elements. Mostly following the notation of [150], let x denote the $1 \times 3k$ column vector of all node coordinates of a given element E, T a tessellation of E into tetrahedra, and $\mathcal{T}_x$ the set of all different tessellations of E. The *mean volume* of E represented by x is given by

$$\text{vol}(x) := \frac{1}{|\mathcal{T}_x|} \sum_{T \in \mathcal{T}_x} \sum_{E^{tet} \in T} \text{vol}(E^{tet}),$$

with $|\mathcal{T}_x|$ denoting the number of different tessellations. Hence, the mean volume of an element is simply the arithmetic mean of the sums of signed tetrahedral volumes for all tessellations.

The denominator $\text{trace}(S_k^t S_k)$ of the mean ratio quality criterion given in (2.6) represents the squared Frobenius norm, which yields the volume scaling invariance while preserving translation and rotation invariance. It is noted in [149] that the volume function restricted to a submanifold incorporating this Frobenius norm is an equivalent way of describing the mean ratio quality measure. This yields a heuristic explanation as to why the global optimization-based approach of Mesquite and GETMe smoothing result in meshes of similar quality, as has been demonstrated by the examples in Chapter 7.

The hexahedron can be tessellated into five and six tetrahedra, respectively. Evaluating $\text{vol}(E^{hex})$ for these tessellations and applying the ∇ operator leads to the associated gradient field $\nabla \text{vol}_{\text{hex}} = (1/12)X_{E^{\text{hex}}}$, where

$$X_{E^{\text{hex}}} := \begin{pmatrix} \nu(p_2,p_5,p_4) + \nu(p_6,p_5,p_8,p_4,p_3,p_2) \\ \nu(p_3,p_6,p_1) + \nu(p_7,p_6,p_5,p_1,p_4,p_3) \\ \nu(p_4,p_7,p_2) + \nu(p_8,p_7,p_6,p_2,p_1,p_4) \\ \nu(p_1,p_8,p_3) + \nu(p_5,p_8,p_7,p_3,p_2,p_1) \\ \nu(p_1,p_6,p_8) + \nu(p_6,p_7,p_8,p_4,p_1,p_2) \\ \nu(p_2,p_7,p_5) + \nu(p_7,p_8,p_5,p_1,p_2,p_3) \\ \nu(p_3,p_8,p_6) + \nu(p_8,p_5,p_6,p_2,p_3,p_4) \\ \nu(p_4,p_5,p_7) + \nu(p_5,p_6,p_7,p_3,p_4,p_1) \end{pmatrix}.$$

The pyramid with quadrilateral base has two different tessellations into two tetrahedra each, which yields $\nabla \text{vol}_{\text{pyr}} = (1/12)X_{E^{\text{pyr}}}$ with

$$X_{E^{\text{pyr}}} := \begin{pmatrix} \nu(p_5,p_4,p_2) + \nu(p_5,p_4,p_3,p_2) \\ \nu(p_5,p_1,p_3) + \nu(p_5,p_1,p_4,p_3) \\ \nu(p_5,p_2,p_4) + \nu(p_5,p_2,p_1,p_4) \\ \nu(p_5,p_3,p_1) + \nu(p_5,p_3,p_2,p_1) \\ 2\nu(p_1,p_2,p_3,p_4) \end{pmatrix}.$$

Finally, there are six different tessellations for the prism. Each is associated to one node of the prism resulting in $\nabla \text{vol}_{\text{pri}} = (1/18)X_{E^{\text{pri}}}$, where

$$X_{E^{\text{pri}}} := \begin{pmatrix} \nu(p_3,p_2,p_4) + \nu(p_2,p_5,p_4,p_6,p_3) \\ \nu(p_1,p_3,p_5) + \nu(p_3,p_6,p_5,p_4,p_1) \\ \nu(p_2,p_1,p_6) + \nu(p_1,p_4,p_6,p_5,p_2) \\ \nu(p_5,p_6,p_1) + \nu(p_6,p_3,p_1,p_2,p_5) \\ \nu(p_6,p_4,p_2) + \nu(p_4,p_1,p_2,p_3,p_6) \\ \nu(p_4,p_5,p_3) + \nu(p_5,p_2,p_3,p_1,p_4) \end{pmatrix}.$$

It should be noticed that the mean volume function is translation and rotation invariant but not scaling invariant. Therefore, by incorporating normalization and translation of the centroid $p_c := (1/k)\sum_{i=1}^{k} p_i$ for an element E with k nodes, an associated quality function is defined by

$$q_v(E) := \text{vol}\left(\frac{1}{\|\tilde{x}_E\|}\tilde{x}_E\right) \quad \text{with} \quad \tilde{x}_E := \left((p_1 - p_c)^{\text{t}},\ldots,(p_k - p_c)^{\text{t}}\right)^{\text{t}}.$$

As is noted in [150], regular tetrahedra, hexahedra, and certain symmetric pyramids and prisms maximize this volume-based quality function. Thus, following the gradient of the volume function by transforming the elements according to (8.7) improves their volume-based quality. By this specific choice of the transformation and quality measure, the given transformations can be analyzed using concepts of global analysis. It is shown that volume elements based on tetrahedra, pyramids, and octahedra are regularized with respect to the mean volume quality if their volume is positive. Furthermore, a volume-based condition is derived under which hexahedral elements are regularized.

Applying the gradient-based transformation iteratively does lead to regular pyramids and prisms with respect to the mean volume criterion, but not with respect to the mean ratio quality criterion. This is because the mean ratio weight matrices given in Figure 2.5 are defined for ideal pyramids and prisms with all edges having the same length, which implies a specific height of these elements. To be compatible to these requirements and to improve regularization speed, [150] proposes alternative transformations by changing the vector fields, which are no longer gradient fields in all cases. Furthermore, an additional element type-specific scaling factor C_{type} is introduced, which is chosen such that the edge lengths after transformation are twice the initial edge lengths of a regular polyhedron for the choice $\sigma = 1$. This results in the new transformation

$$x' := x + \sigma C_{\text{type}} \Psi(\widetilde{X}_{E\text{type}}),$$

with an element type-specific vector field $\widetilde{X}_{E\text{type}}$.

The modified transformation of a tetrahedron uses the same vector field as given in (8.6). However, to conform to the notation of the subsequently introduced vector fields and for completeness, it is described as follows. Using the auxiliary expressions

$$\nu_1 := \nu(p_4, p_3, p_2), \quad \nu_2 := \nu(p_4, p_1, p_3), \quad \text{and} \quad \nu_3 := \nu(p_4, p_2, p_1),$$

the tetrahedron transformation vector field is defined as

$$\widetilde{X}_{E\text{tet}} := (\nu_1^t, \nu_2^t, \nu_3^t, -\nu_1^t - \nu_2^t - \nu_3^t)^t.$$

By solving the edge length equation $\|p_1' - p_2'\| = 2\|p_1 - p_2\|$, it follows that the associated scaling factor is given by $C_{\text{tet}} = \sqrt[4]{3}/\sqrt{2}$. In total, transforming a tetrahedron requires the evaluation of three different cross products.

The vector field for transforming hexahedral elements is defined with the aid of the expressions

$$\nu_4 := \nu(p_1, p_5, p_8, p_4), \quad \nu_5 := \nu(p_1, p_2, p_6, p_5), \quad \nu_6 := \nu(p_2, p_3, p_7, p_6),$$
$$\nu_7 := \nu(p_3, p_4, p_8, p_7), \quad \nu_8 := \nu(p_1, p_4, p_3, p_2)$$

and their linear combinations

$$\lambda_1 := \nu_4 + \nu_5 + \nu_8, \qquad \lambda_2 := \nu_5 + \nu_6 + \nu_8,$$
$$\lambda_3 := \nu_6 + \nu_7 + \nu_8, \qquad \lambda_4 := \nu_7 + \nu_4 + \nu_8,$$

as

$$\widetilde{X}_{E\text{hex}} := (\lambda_1^t, \lambda_2^t, \lambda_3^t, \lambda_4^t, -\lambda_3^t, -\lambda_4^t, -\lambda_1^t, -\lambda_3^t)^t,$$

which requires the computation of five different cross products. The scaling factor results in $C_{\text{hex}} = \sqrt[4]{6}/2$.

Transforming a pyramid with a quadrilateral base also requires the computation of five cross products for its entries based on linear combinations of

$$\nu_9 := \nu(p_5, p_4, p_3, p_2), \quad \nu_{10} := \nu(p_5, p_1, p_4, p_3), \quad \nu_{11} := \nu(p_1, p_2, p_3, p_4)$$
$$\nu_{12} := \nu(p_5, p_4, p_2), \qquad \nu_{13} := \nu(p_5, p_1, p_3).$$

The vector field for the flow-based regularizing pyramid transformation with respect to the mean ratio quality criterion is given by

$$
\tilde{X}_{E_{\mathrm{pyr}}} := \begin{pmatrix} \nu_9 + 4\nu_{12} \\ \nu_{10} + 4\nu_{13} \\ -\nu_{11} - \nu_9 - 4\nu_{12} \\ -\nu_{11} - \nu_{10} - 4\nu_{13} \\ 2\nu_{11} \end{pmatrix},
$$

with the element type-specific scaling factor $C_{\mathrm{pyr}} = \sqrt[4]{30}/5$.

Finally, using the auxiliary expressions

$$
\begin{aligned}
&\nu_{14} := \nu(p_2, p_5, p_6, p_3), \quad \nu_{15} := \nu(p_3, p_6, p_4, p_1), \quad \nu_{16} := \nu(p_1, p_4, p_5, p_2), \\
&\nu_{17} := \nu(p_5, p_4, p_6), \qquad\quad \nu_{18} := \nu(p_1, p_2, p_3),
\end{aligned}
$$

the prism with triangular base is regularized by the vector field

$$
\tilde{X}_{E_{\mathrm{pri}}} := \begin{pmatrix} \nu_{14} + 2\nu_{17} \\ \nu_{15} + 2\nu_{17} \\ \nu_{16} + 2\nu_{17} \\ \nu_{14} + 2\nu_{18} \\ \nu_{15} + 2\nu_{18} \\ \nu_{16} + 2\nu_{18} \end{pmatrix}
$$

and the associated scaling factor $C_{\mathrm{pri}} = \sqrt[4]{378}/6$. This representation also involves the computation of five different cross products.

To give a comparison of the rates of regularization of these flow-based transformations with those of the dual element-based polyhedra transformations, the same numerical test has been conducted as is described at the end of Section 5.2.2. That is, for 100,000 initial elements E_j of a given type with random node positions and 101 equidistant values $\sigma_k \in [0, 2]$, the flow-based transformation was iteratively applied to each combination (E_j, σ_k) until the resulting element mean ratio quality number deviated less than 10^{-6} from the regular element mean ratio quality number one.

As described in Section 5.2.2, depending on the element type, random initial elements are invalid in the majority of cases. However, element validity is no prerequisite for resulting in regular elements after iteratively applying the flow-based transformation. Taking the arithmetic mean of all 100,000 iteration numbers obtained to reach the quality limit for a specific choice of σ_k results in the convergence graph depicted in Figure 8.6a. For specific choices of σ, the standard deviation is given by error markers. Results for randomly chosen valid initial elements with respect to the mean ratio quality criterion are shown in Figure 8.6b.

It can be seen that the speed of regularization is similar for hexahedral, pyramidal, and prismatic elements for arbitrary initial elements as well as valid initial elements. Compared to this, the average iteration number required to regularize tetrahedral elements is significantly lower, which should

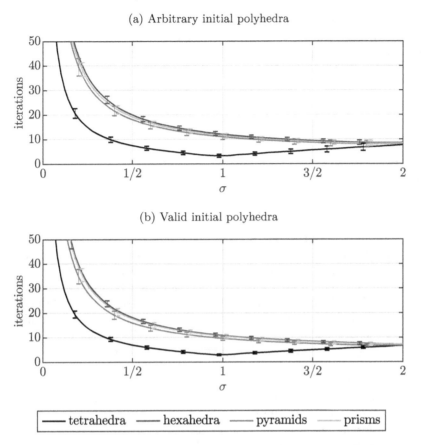

Figure 8.6: Average iteration numbers with respect to transformation parameter $\sigma \in [0, 2]$, required to regularize random polyhedra by flow-based transformations.

be compensated in GETMe smoothing by element type-specific choices of the transformation parameter σ. By comparing the upper graphs for arbitrary initial elements with the lower graphs for valid initial elements, it can also be seen that element validity has a minor impact on the rate of regularization. This is also reflected by the slightly increased standard deviation for arbitrary initial elements.

By conducting the same test as in the case of the dual element-based transformations, results are directly comparable to those depicted in Figure 5.24. Both transformation types achieve comparable rates of regularization, though for different values of σ. In addition, both transformation schemes have a preferred element type, for which, on average, elements are transformed in fewer iterations as compared to the other element types. In the case of the flow-based transformations, this is the tetrahedron, in the case of the dual element-based transformations it is the hexahedron. By comparing the error markers, it can also be deduced that the standard deviation of the iteration numbers required to regularize elements by the given flow-based transformations is smaller if compared to those of the dual element-based transformations. Nevertheless, both transformations lead to comparable results if used by GETMe smoothing [150].

8.2 Variations of GETMe smoothing

8.2.1 Geometric triangle transformation-based smoothing

From a mesh smoothing point of view, the element transformations for tetrahedral and hexahedral elements based on triangle transformations as described in Section 8.1.1 are already mesh smoothing algorithms, since the faces of the original or dual element are transformed and new node positions are derived by node averaging. To speed up the regularizing effect of the transformation, [147] proposes to apply the triangle transformation repeatedly a given number of times to the triangular faces before node averaging is conducted.

As can be seen in Figure 8.4, applying the transformation can change the element size significantly. Therefore, as in the case of the GETMe approach proposed in Chapter 6, elements are rescaled after transformation. This can, for example, be based on preserving the initial element area or volume.

Since the GETMe simultaneous scheme given in Section 6.1.1 utilizes element transformations as building blocks, it can be applied without modification. That is, mesh smoothing for the given triangle-based element transformation is conducted by applying the following steps:

1. Transform each element individually by applying the triangle transformation a prescribed number of times.

2. Rescale each element to preserve its initial area or volume.

3. Determine new mesh nodes as the arithmetic mean of transformed element nodes.

4. Reset fixed boundary nodes and nodes of inverted elements.

5. Compute mesh quality and terminate if improvement is below a given threshold.

It is remarked in [147] that a hexahedral mesh could alternatively be smoothed by subdividing each hexahedron into four tetrahedra whose edges are given by the diagonals of the faces of the hexahedron. Similar to the quadrilateral transformation, element transformation is then based on transforming the four tetrahedra. The same idea applies to all platonic solids, which can be subdivided into regular tetrahedra.

Like the relaxation parameter in GETMe and GETMe adaptive smoothing, the GETMe variants proposed in [147] introduce parameters for adaptive smoothing control, which further improve the resulting mesh quality. This is demonstrated by a variety of numerical examples.

8.2.2 GETOpt

A new global optimization-based smoothing approach for tetrahedral meshes based on the ideas of GETMe smoothing and flow-based transformations is presented in [148]. Here, a global mesh quality metric is introduced as an objective function, which also incorporates size of elements and whose gradient flow yields a combination of GETMe and Laplacian smoothing. Since applying this method tends to result in tetrahedra of comparable size, it is primarily intended for isotropic mesh smoothing. The resulting method, named GETOpt, is described briefly in the following.

Considering a standard gradient flow-based continuous smoothing scheme for a tetrahedral mesh with the node, element, and edge sets $\mathcal{V}$, $\mathcal{T}$, and $\mathcal{E}$, respectively, the key ingredient of smoothing is the quality measure $m : \mathcal{T} \to \mathbb{R}$, serving as objective function. Each node $p \in \mathcal{V}$ evolves according to $dp/dt = \partial m/\partial p$. For example, taking the sum of all squared edge lengths as quality measure, i.e.,

$$m_{\text{length}} := \sum_{e \in \mathcal{E}} \text{length}(e)^2,$$

leads to an infinitesimal version of Laplacian smoothing [148]. In contrast, using the arithmetic mean of the logarithms of all tetrahedron volumes in the mesh, i.e.,

$$m_{\text{vol}} := \log V := \frac{1}{|\mathcal{T}|} \sum_{T \in \mathcal{T}} \log \text{vol}(T),$$

as quality measure m results in a scaled sum of opposite face normals for dp/dt. That is, the gradient flow for m_{vol} is related to GETMe smoothing using the opposite face normals tetrahedron transformation given in Section 5.2.1. However, GETMe smoothing differs by its element-oriented scaling schemes.

By combining the length and the volume-based quality metrics, the resulting global optimization scheme, called *GETOpt*, is based on the global mesh quality measure

$$m := m_{\text{vol}} - \frac{3}{2} \log m_{\text{length}},$$

which combines the two aforementioned quality metrics. This mesh quality measure is scaling invariant and not related to a scaling invariant element quality measure. It is shown in [148] that by following the gradient flow of m, no individual tetrahedron can degenerate and the diameter of the largest tetrahedron is bounded by some multiple of the diameter of the smallest tetrahedron.

Let $L := \sqrt{m_{\text{length}}}$, i.e., L is the square root of the sum of all squared edge lengths. It follows from the standard results for gradient flows that the flow must converge to a critical point of the function $\log V - 3 \log L$, which could be a local maximum or saddle point. However, it is stated in [148] that the latter does not happen for generic initial data.

Let $n(p, T)$ denote the outward directed normal to the face of T opposite an arbitrary mesh node p. Here, the length of $n(p, T)$ is twice the area of that face. Discretization of dp/dt for this scheme leads to

$$p' := p + \frac{\delta L^2}{6|\mathcal{T}|} \sum_T \frac{1}{\mathrm{vol}(T)} n(p, T) + 3\delta \sum_q (q - p) \qquad (8.8)$$

for the new node position p' of p after one smoothing step. In (8.8) the first sum is taken for all tetrahedra T attached to the node p and the second sum is over all nodes q that are edge-connected to p. Furthermore, δ is a scaling invariant discretization parameter, which should be chosen according to

$$\delta \ll \frac{|\mathcal{T}|}{L^2} \min_{p,T} \frac{\mathrm{vol}(T)^2}{|n(p, T)|^2},$$

i.e., it should be chosen small compared to the minimum height of a tetrahedron in the entire mesh.

In [148] the second variation of $\log V - 3 \log L$ is derived. Since it has an unclear sign, it cannot be concluded that every critical point is a local maximum. However, the effectivity of GETOpt-based mesh smoothing with respect to resulting mesh quality and finite element solution accuracy is shown by a broad variety of numerical tests based on the examples given in [162, 163]. Results are compared to those of other global optimization-based methods implemented by Mesquite. Although GETOpt is not geared towards improving the mean ratio quality, it results in comparable mesh quality numbers. Furthermore, GETOpt results in slightly better finite element solution accuracy and, due to construction, conforms the element volumes of the mesh better than Mesquite.

Since GETMe is related to the approach of GETOpt, these results also give an indication why GETMe smoothing yields meshes of high quality comparable to those of global optimization-based methods. Here, GETMe smoothing is computationally less expensive if compared to GETOpt smoothing.

Bibliography

[1] AIM@SHAPE Mesh Repository. `http://shapes.aim-at-shape.net/`, accessed July 17, 2012.

[2] M. Ainsworth and J.T. Oden. *A Posteriori Error Estimation in Finite Element Analysis*. John Wiley & Sons, 2000.

[3] N. Amenta, M. Bern, and D. Eppstein. Optimal point placement for mesh smoothing. *Journal of Algorithms*, 30(2):302–322, 1999.

[4] J.H. Argyris. *Energy theorems and structural analysis*. Butterworth, 1960.

[5] J.H. Argyris. Personal communication, 1995.

[6] I. Babuška and A.K. Aziz. On the angle condition in the finite element method. *SIAM Journal on Numerical Analysis*, 13(2):214–226, 1976.

[7] I. Babuška, U. Banerjee, and J.E. Osborn. Survey of meshless and generalized finite element methods: a unified approach. *Acta Numerica*, 12:1–125, 2003.

[8] B. Balendran. A direct smoothing method for surface meshes. In *Proceedings of the 8th International Meshing Roundtable*, pages 189–193, 1999.

[9] R.E. Bank and R.K. Smith. Mesh smoothing using a posteriori error estimates. *SIAM Journal on Numerical Analysis*, 34(3):979–997, 1997.

[10] S.E. Benzley, E. Perry, K. Merkley, B. Clark, and G. Sjaardema. A comparison of all hexagonal and all tetrahedral finite element meshes for elastic and elastic-plastic analysis. In *Proceedings of the 4th International Meshing Roundtable*, pages 179–191, 1995.

[11] M. Bern, D. Eppstein, and J. Gilbert. Provably good mesh generation. *Journal of Computer and System Sciences*, 48(3):384–409, 1994.

[12] M. Berzins. Mesh quality: a function of geometry, error estimates or both? *Engineering with Computers*, 15(3):236–247, 1999.

[13] A. Bowyer. Computing Dirichlet tessellations. *The Computer Journal*, 24(2):162–166, 1981.

[14] D. Braess. *Finite Elemente.* Springer-Verlag, 4th edition, 2007.

[15] S.C. Brenner and L.R. Scott. *The Mathematical Theory of Finite Element Methods.* Number 15 in Texts in Applied Mathematics. Springer-Verlag, 3rd edition, 2008.

[16] M. Brewer, L.A. Freitag Diachin, P.M. Knupp, T. Leurent, and D. Melander. The Mesquite mesh quality improvement toolkit. In *Proceedings of the 12th International Meshing Roundtable*, pages 239–250, 2003.

[17] I.N. Bronštein, K.A. Semendjajew, G. Musiol, and H. Mühlig. *Taschenbuch der Mathematik.* Verlag Harri Deutsch, 4th edition, 2000.

[18] V.I. Burenkov. *Sobolev Spaces on Domains*, volume 137 of *Teubner-Texte zur Mathematik.* B.G. Teubner, 1998.

[19] N.A. Calvo and S.R. Idelsohn. All-hexahedral mesh smoothing with a node-based measure of quality. *International Journal for Numerical Methods in Engineering*, 50(8):1957–1967, 2001.

[20] N.A. Calvo, S.R. Idelsohn, and E. Oñate. The extended Delaunay tessellation. *Engineering Computations: International Journal for Computer-Aided Engineering*, 20(5/6):583–600, 2003.

[21] S.A. Canann, J.R. Tristano, and M.L. Staten. An approach to combined Laplacian and optimization-based smoothing for triangular, quadrilateral, and quad-dominant meshes. In *Proceedings of the 7th International Meshing Roundtable*, pages 479–494, 1998.

[22] L. Chen. Mesh smoothing schemes based on optimal Delaunay triangulations. In *Proceedings of the 13th International Meshing Roundtable*, pages 109–120, 2004.

[23] Z. Chen, J.R. Tristano, and W. Kwok. Combined Laplacian and optimization-based smoothing for quadratic mixed surface meshes. In *Proceedings of the 12th International Meshing Roundtable*, pages 201–213, 2003.

[24] A.O. Cifuentes and A. Kalbag. A performance study of tetrahedral and hexahedral elements in 3-d finite element structural analysis. *Finite Elements in Analysis and Design*, 12(3–4):313–318, 1992.

[25] R.W. Clough. Early history of the finite element method from the view point of a pioneer. *International Journal for Numerical Methods in Engineering*, 60(1):283–287, 2004.

[26] R.W. Clough. The finite element method in plane stress analysis. In *Proceedings of the 2nd ASCE Conference on Electronic Computation*, pages 345–378, 1960.

[27] R. Courant. Variational methods for the solution of problems of equilibrium and vibrations. *Bulletin of the American Mathematical Society*, 49(1):1–23, 1943.

[28] F. Courty, D. Leservoisier, P.L. George, and A. Dervieux. Continuous metrics and mesh adaptation. *Applied Numerical Mathematics*, 56(2):117–145, 2006.

[29] P.J. Davis. *Circulant matrices*. Chelsea Publishing, 2nd edition, 1994.

[30] B. Delaunay. Sur la sphère vide. *Bulletin de l'Académie des Sciences de l'URSS*, 7(6):793–800, 1934.

[31] M.F.F. Delgado. Entwicklung und Erprobung eines Generators für Volumenelemente. Master's thesis, Institut für Statik und Dynamik der Luft- und Raumfahrtkonstruktionen, University of Stuttgart, Germany, 1988.

[32] D. Delso. Munich Olympic Park image. Wikimedia Commons, https://commons.wikimedia.org/wiki/File:Olympiastadion,_M%C3%BCAnich,_Alemania_2012-04-28,_DD_03.JPG, Licence: CC-BY-SA, accessed July 30, 2018.

[33] A.R. Diaz, N. Kikuchi, and J.E. Taylor. A method of grid optimization for finite element methods. *Computer Methods in Applied Mechanics and Engineering*, 41(1):29–45, 1983.

[34] J. Douglas. On linear polygon transformations. *Bulletin of the American Mathematical Society*, 46(6):551–560, 1940.

[35] Drexel University, Geometric & Intelligent Computing Laboratory model repository. http://edge.cs.drexel.edu/repository/, accessed July 3, 2012.

[36] Q. Du and D. Wang. Tetrahedral mesh generation and optimization based on centroidal Voronoi tessellations. *International Journal for Numerical Methods in Engineering*, 56(9):1355–1373, 2003.

[37] R.A. Dwyer. A faster divide-and-conquer algorithm for constructing Delaunay triangulations. *Algorithmica*, 2(1–4):137–151, 1987.

[38] V. Dyedov, D.R. Einstein, X. Jiao, A.P. Kuprat, J.P. Carson, and F. del Pin. Variational generation of prismatic boundary-layer meshes for biomedical computing. *International Journal for Numerical Methods in Engineering*, 79(8):907–945, 2009.

[39] H. Edelsbrunner. *Geometry and Topology for Mesh Generation*. Cambridge University Press, 2001.

[40] H. Edelsbrunner, T.S. Tan, and R. Waupotitsch. An $O(n^2 \log n)$ time algorithm for the minmax angle triangulation. *SIAM Journal on Scientific and Statistical Computing*, 13(4):994–1008, 1992.

[41] O. Egorova, M. Savchenko, N. Kojekine, I. Semonova, I. Hagiwara, and V. Savchenko. Improvement of mesh quality using a statistical approach. In *Proceedings of the 3rd IASTED international conference on visualization, imaging, and image processing*, volume 2, pages 1016–1021, 2003.

[42] A. El-Hamalawi. A 2D combined advancing front-Delaunay mesh generation scheme. *Finite Elements in Analysis and Design*, 40(9–10):967–989, 2004.

[43] J.M. Escobar, R. Montenegro, G. Montero, E. Rodríguez, and J.M. González-Yuste. Smoothing and local refinement techniques for improving tetrahedral mesh quality. *Computers & Structures*, 83(28–30):2423–2430, 2005.

[44] J.M. Escobar, G. Montero, R. Montenegro, and E. Rodríguez. An algebraic method for smoothing surface triangulations on a local parametric space. *International Journal for Numerical Methods in Engineering*, 66(4):740–760, 2006.

[45] D.A. Field. Laplacian smoothing and Delaunay triangulations. *Communications in Applied Numerical Methods*, 4(6):709–712, 1988.

[46] N. Flandrin, H. Borouchaki, and C. Bennis. 3D hybrid mesh generation for reservoir simulation. *International Journal for Numerical Methods in Engineering*, 65(10):1639–1672, 2006.

[47] N.T. Folwell and S.A. Mitchell. Reliable whisker weaving via curve contraction. *Engineering with Computers*, 15(3):292–302, 1999.

[48] L.A. Freitag. On combining Laplacian and optimization-based mesh smoothing techniques. *AMD Trends in Unstructured Mesh Generation*, 220:37–43, 1997.

[49] L.A. Freitag, M. Jones, and P. Plassmann. A parallel algorithm for mesh smoothing. *SIAM Journal on Scientific Computing*, 20(6):2023–2040, 1999.

[50] L.A. Freitag and C. Ollivier-Gooch. Tetrahedral mesh improvement using swapping and smoothing. *International Journal for Numerical Methods in Engineering*, 40(21):3979–4002, 1997.

[51] L.A. Freitag and C. Ollivier-Gooch. A cost/benefit analysis of simplicial mesh improvement techniques as measured by solution efficiency. *International Journal of Computational Geometry & Applications*, 10(4):361–382, 2000.

[52] L.A. Freitag Diachin, P.M. Knupp, T. Munson, and S.M. Shontz. A comparison of inexact Newton and coordinate descent mesh optimization techniques. In *Proceedings of the 13th International Meshing Roundtable*, pages 243–254, 2004.

[53] L.A. Freitag Diachin, P.M. Knupp, T. Munson, and S.M. Shontz. A comparison of two optimization methods for mesh quality improvement. *Engineering with Computers*, 22(2):61–74, 2006.

[54] P.J. Frey and L. Marechal. Fast adaptive quadtree mesh generation. In *Proceedings of the 7th International Meshing Roundtable*, pages 211–224, 1998.

[55] P.J. Frey and P.L. George. *Mesh Generation: Application to Finite Elements*. Wiley-ISTE, 2nd edition, 2008.

[56] Gamma project mesh database. `http://www-roc.inria.fr/gamma/download/`, accessed September 9, 2010.

[57] R.V. Garimella, M.J. Shashkov, and P.M. Knupp. Triangular and quadrilateral surface mesh quality optimization using local parametrization. *Computer Methods in Applied Mechanics and Engineering*, 193(9–11):913–928, 2004.

[58] C.F. Gauß. *Disquisitiones Arithmeticae*. Gerhard Fleischer, 1801.

[59] GetFEM++: An open-source finite element library, version 4.3. `http://getfem.org/`, accessed September 16, 2018.

[60] D. Gilbarg and N.S. Trudinger. *Elliptic Partial Differential Equations of Second Order*. Springer-Verlag, 2001.

[61] Google Scholar. `https://scholar.google.de/`, accessed November 5, 2017.

[62] R.M. Gray. Toeplitz and circulant matrices: A review. *Foundations and Trends in Communications and Information Theory*, 2(3):155–239, 2006.

[63] A. Greenbaum. *Iterative Methods for Solving Linear Systems*. Number 17 in Frontiers in Applied Mathematics. SIAM, 1997.

[64] K.K. Gupta and J.L. Meek. A brief history of the beginning of the finite element method. *International Journal for Numerical Methods in Engineering*, 39(22):3761–3774, 1996.

[65] G. Hansen, A. Zardecki, D. Greening, and R. Bos. A finite element method for three-dimensional unstructured grid smoothing. *Journal of Computational Physics*, 202(1):281–297, 2005.

[66] J. Heuser. Aorta scheme image. Wikimedia Commons, `https://commons.wikimedia.org/wiki/File:Aorta_scheme.jpg`, Licence: CC-BY-SA-3.0 / GFDL, accessed July 30, 2018.

[67] K. Höllig. *Finite Element Methods with B-Splines.* Number 26 in Frontiers in Applied Mathematics. SIAM, 2003.

[68] K. Höllig, U. Reif, and J. Wipper. B-spline approximation of Neumann problems. Preprint 2001-2, Mathematisches Institut A, Universität Stuttgart, 2001.

[69] K. Höllig, U. Reif, and J. Wipper. Error estimates for the WEB-spline method. In T. Lyche and L.L. Schumaker, editors, *Mathematical Methods for Curves and Surfaces: Oslo 2000*, pages 195–209. Vanderbilt University Press, 2001.

[70] K. Höllig, U. Reif, and J. Wipper. Weighted extended B-spline approximation of Dirichlet problems. *SIAM Journal on Numerical Analysis*, 39(2):442–462, 2001.

[71] K. Höllig, U. Reif, and J. Wipper. Multigrid methods with WEB-splines. *Numerische Mathematik*, 91(2):237–255, 2002.

[72] K. Höllig, U. Reif, and J. Wipper. Verfahren zur Erhöhung der Leistungsfähigkeit einer Computereinrichtung bei Finite-Elemente-Simulationen und eine solche Computereinrichtung. German patent, publication number DE10023377C2, 2003.

[73] K.H. Huebner, D.L. Dewhirst, D.E. Smith, and T.G. Byrom. *The Finite Element Method for Engineers.* John Wiley & Sons, 4th edition, 2001.

[74] Y. Ito and K. Nakahashi. Improvements in the reliability and quality of unstructured hybrid mesh generation. *International Journal for Numerical Methods in Fluids*, 45(1):79–108, 2004.

[75] Y. Ito, A.M. Shih, and B.K. Soni. Reliable isotropic tetrahedral mesh generation based on an advancing front method. In *Proceedings of the 13th International Meshing Roundtable*, pages 95–106, 2004.

[76] Y. Ito, A.M. Shih, and B.K. Soni. Octree-based reasonable-quality hexahedral mesh generation using a new set of refinement templates. *International Journal for Numerical Methods in Engineering*, 77(13):1809–1833, 2009.

[77] Jebulon. Gorgon medusa floor mosaic image. Wikimedia Commons, `https://commons.wikimedia.org/wiki/File:Mosaic_floor_opus_tessellatum_detail_Gorgone_NAMA_Athens_Greece.jpg`, Licence: CC0 1.0, accessed July 30, 2018.

[78] X. Jiao, D. Wang, and H. Zha. Simple and effective variational optimization of surface and volume triangulations. In *Proceedings of the 17th International Meshing Roundtable*, pages 315–332, 2008.

[79] R.E. Jones. A self-organizing mesh generation program. *Journal of Pressure Vessel Technology*, 96(3):193–199, 1974.

[80] B.W. Kernighan and D.M. Ritchie. *The C Programming Language*. Prentice Hall, 2nd edition, 1988.

[81] B.M. Klingner and J.R. Shewchuk. Aggressive tetrahedral mesh improvement. In *Proceedings of the 16th International Meshing Roundtable*, pages 3–23, 2007.

[82] P.M. Knupp. Next-generation sweep tool: A method for generating all-hex meshes on two-and-one-half dimensional geometries. In *Proceedings of the 7th International Meshing Roundtable*, pages 505–513, 1998.

[83] P.M. Knupp. Hexahedral mesh untangling & algebraic mesh quality metrics. In *Proceedings of the 9th International Meshing Roundtable*, pages 173–183, 2000.

[84] P.M. Knupp. Algebraic mesh quality metrics. *SIAM Journal on Scientific Computing*, 23(1):193–218, 2001.

[85] P.M. Knupp. Remarks on mesh quality. In *Proceedings of the 45th AIAA Aerospace Sciences Meeting and Exhibit*, 2007.

[86] P.M. Knupp. Introducing the target-matrix paradigm for mesh optimization via node-movement. In *Proceedings of the 19th International Meshing Roundtable*, pages 67–84. Springer, 2010.

[87] S. Kulovec, L. Kos, and J. Duhovnik. Mesh smoothing with global optimization under constraints. *Strojniški vestnik - Journal of Mechanical Engineering*, 57(7–8):555–567, 2011.

[88] D.T. Lee and B.J. Schachter. Two algorithms for constructing a Delaunay triangulation. *International Journal of Computer and Information Sciences*, 9(3):219–242, 1980.

[89] T.S. Li, R.M. McKeag, and C.G. Armstrong. Hexahedral meshing using midpoint subdivision and integer programming. *Computer Methods in Applied Mechanics and Engineering*, 124(1–2):171–193, 1995.

[90] X.Y. Li and L.A. Freitag. Optimization-based quadrilateral and hexahedral mesh untangling and smoothing techniques. Technical report, Argonne National Laboratory, 1999.

[91] V.D. Liseikin. *Grid Generation Methods*. Springer, 3rd edition, 2017.

[92] G.R. Liu. An overview on meshfree methods: for computational solid mechanics. *International Journal of Computational Methods*, 13(5):1630001, 2016.

[93] D.S.H. Lo. *Finite Element Mesh Generation*. CRC Press, 2017.

[94] S.H. Lo. A new mesh generation scheme for arbitrary planar domains. *International Journal for Numerical Methods in Engineering*, 21(8):1403–1426, 1985.

[95] R. Löhner, K. Morgan, and O.C. Zienkiewicz. Adaptive grid refinement for the compressible Euler equations. In I. Babuška, O.C. Zienkiewicz, J. Gago, and E.R. deA. Oliveira, editors, *Accuracy estimates and adaptive refinements in finite element computations*, pages 281–297. John Wiley & Sons, 1986.

[96] R. Löhner and P. Parikh. Generation of three-dimensional unstructured grids by the advancing-front method. *International Journal for Numerical Methods in Fluids*, 8(10):1135–1149, 1988.

[97] MathWorks Inc. MATLAB R2018a. http://www.mathworks.com/, accessed April 2, 2018.

[98] A. Menéndez-Díaz, C. González-Nicieza, and A.E. Álvarez-Vigil. Hexahedral mesh smoothing using a direct method. *Computers & Geosciences*, 31(4):453–463, 2005.

[99] Mesquite: mesh quality improvement toolkit. https://trilinos.org/oldsite/packages/mesquite/, accessed September 22, 2018.

[100] T. Munson. Mesh shape-quality optimization using the inverse mean-ratio metric. *Mathematical Programming*, 110(3):561–590, 2007.

[101] B.H. Neumann. Some remarks on polygons. *Journal of the London Mathematical Society*, s1-16(4):230–245, 1941.

[102] V.P. Nguyen, T. Rabczuk, S. Bordas, and M. Duflot. Meshless methods: A review and computer implementation aspects. *Mathematics and Computers in Simulation*, 79(3):763–813, 2008.

[103] OpenMP Architecture Review Board. *OpenMP Application Program Interface, Version 3.1*, 2011.

[104] S.J. Owen and M.S. Shephard, editors. *International Journal for Numerical Methods in Engineering*, Special Issue: Trends in Unstructured Mesh Generation, volume 58, 2003.

[105] S.J. Owen. A survey of unstructured mesh generation technology. In *Proceedings of the 7th International Meshing Roundtable*, pages 239–267, 1998.

[106] C.C. Pain, A.P. Umpleby, C.R.E. de Oliveira, and A.J.H. Goddard. Tetrahedral mesh optimisation and adaptivity for steady-state and transient finite element calculations. *Computer Methods in Applied Mechanics and Engineering*, 190(29–30):3771–3796, 2001.

[107] M. Papadrakakis, editor. *Solving large-scale problems in mechanics. The development and application of computational solution methods*. John Wiley & Sons, 1993.

[108] K. Petr. Ein Satz über Vielecke. *Archiv der Mathematik und Physik: mit besonderer Rücksicht auf die Bedürfnisse der Lehrer an höheren Unterrichtsanstalten*, 13:29–31, 1908.

[109] M.A. Price and C.G. Armstrong. Hexahedral mesh generation by medial surface subdivision: Part II. solids with flat and concave edges. *International Journal for Numerical Methods in Engineering*, 40(1):111–136, 1997.

[110] J.W.S. Rayleigh. *The Theory of Sound, Volume Two*. Dover Publications, 2nd edition, 1896.

[111] F.T. Reusch. Entwicklung einer dynamischen Programmstruktur für einen 3D-Volumengenerator. Master's thesis, Institut für Statik und Dynamik der Luft- und Raumfahrtkonstruktionen, Universität Stuttgart, 1988.

[112] W. Ritz. Über eine neue Methode zur Lösung gewisser Variationsprobleme der mathematischen Physik. *Journal für die reine und angewandte Mathematik*, 135:1–61, 1909.

[113] M.C. Rivara. New longest-edge algorithms for the refinement and/or improvement of unstructured triangulations. *International Journal for Numerical Methods in Engineering*, 40(18):3313–3324, 1997.

[114] X. Roca, J. Sarrate, and A. Huerta. Surface mesh projection for hexahedral mesh generation by sweeping. In *Proceedings of the 13th International Meshing Roundtable*, pages 169–180, 2004.

[115] D. Rypl. Approaches to discretization of 3D surfaces. In *CTU Reports*, volume 7. CTU Publishing House, 2003.

[116] A. Samuelsson and O.C. Zienkiewicz. History of the stiffness method. *International Journal for Numerical Methods in Engineering*, 67(2):149–157, 2006.

[117] M. Savchenko, O. Egorova, I. Hagiwara, and V. Savchenko. An approach to improving triangular surface mesh. *JSME International Journal, Series C, Mechanical Systems, Machine Elements and Manufacturing*, 48(2):137–148, 2005.

[118] M. Savchenko, O. Egorova, I. Hagiwara, and V. Savchenko. Hexahedral mesh improvement algorithm. *JSME International Journal, Series C, Mechanical Systems, Machine Elements and Manufacturing*, 48(2):130–136, 2005.

[119] V. Savchenko, M. Savchenko, O. Egorova, and I. Hagiwara. Mesh quality improvement: Radial basis functions approach. *International Journal of Computer Mathematics*, 85(10):1589–1607, 2008.

[120] I. Sazonov, D. Wang, O. Hassan, K. Morgan, and N. Weatherill. A stitching method for the generation of unstructured meshes for use with co-volume solution techniques. *Computer Methods in Applied Mechanics and Engineering*, 195(13–16):1826–1845, 2006.

[121] M.S. Shephard and M.K. Georges. Automatic three-dimensional mesh generation by the finite octree technique. *International Journal for Numerical Methods in Engineering*, 32(4):709–749, 1991.

[122] J. Shepherd, S.A. Mitchell, P.M. Knupp, and D. White. Methods for multisweep automation. In *Proceedings of the 9th International Meshing Roundtable*, pages 77–87, 2000.

[123] J.R. Shewchuk. Tetrahedral mesh generation by Delaunay refinement. In *Proceedings of the 14th Annual Symposium on Computational Geometry*, pages 86–95, 1998.

[124] J.R. Shewchuk. Lecture notes on Delaunay mesh generation. Technical report, Department of Electrical Engineering and Computer Sciences, University of California at Berkeley, 1999.

[125] J.R. Shewchuk. What is a good linear element? Interpolation, conditioning, and quality measures. In *Proceedings of the 11th International Meshing Roundtable*, pages 115–126, 2002.

[126] K. Shimada and D.C. Gossard. Bubble mesh: automated triangular meshing of non-manifold geometry by sphere packing. In *Proceedings of the 3rd ACM symposium on solid modeling and applications*, pages 409–419, 1995.

[127] R.B. Simpson. Geometry independence for a meshing engine for 2D manifolds. *International Journal for Numerical Methods in Engineering*, 60(3):675–694, 2004.

[128] Y. Sirois, J. Dompierre, M.G. Vallet, and F. Guibault. Hybrid mesh smoothing based on Riemannian metric non-conformity minimization. *Finite Elements in Analysis and Design*, 46(1–2):47–60, 2010.

[129] M.L. Staten, R.A. Kerr, S.J. Owen, and T.D. Blacker. Unconstrained paving and plastering: Progress update. In *Proceedings of the 15th International Meshing Roundtable*, pages 469–486, 2006.

[130] J. Stoer. *Numerische Mathematik 1.* Springer-Verlag, 6th edition, 1993.

[131] G. Strang and G. Fix. *An Analysis of the Finite Element Method.* Wellesley-Cambridge Press, 2nd edition, 2008.

[132] B. Stroustrup. *The C++ Programming Language.* Addison-Wesley, 4th edition, 2013.

[133] T.J. Tautges. The generation of hexahedral meshes for assembly geometry: survey and progress. *International Journal for Numerical Methods in Engineering*, 50(12):2617–2642, 2001.

[134] T.J. Tautges, T. Blacker, and S.A. Mitchell. The whisker weaving algorithm: A connectivity-based method for constructing all-hexahedral finite element meshes. *International Journal for Numerical Methods in Engineering*, 39(19):3327–3349, 1996.

[135] L.T. Tenek and J.H. Argyris. *Finite element analysis for composite structures.* Kluwer Academic Publishers, 1998.

[136] J.F. Thompson, B.K. Soni, and N.P. Weatherill, editors. *Handbook of grid generation.* CRC Press, 1998.

[137] M.J. Turner, R.W. Clough, H.C. Martin, and L.J. Topp. Stiffness and deflection analysis of complex structures. *Journal of the Aeronautical Sciences*, 23(9):805–823, 1956.

[138] TWT. Aletis vehicle design and prototyping: Aerodynamic analysis. Technical Report 04/04, TWT GmbH, 2001.

[139] D. Vartziotis. Articulate endoprosthesis for total hip arthroplasty (in Greek). Greek patent, application number GR20050100072 (A), 2006.

[140] D. Vartziotis. Problem 11328. *American Mathematical Monthly*, 114(10):925, 2007.

[141] D. Vartziotis. *A finite element mesh smoothing method based on geometric element transformations.* PhD thesis, National Technical University of Athens, 2013. http://thesis.ekt.gr/thesisBookReader/id/38257.

[142] D. Vartziotis. Problem 11860. *American Mathematical Monthly*, 122(8):801, 2015.

[143] D. Vartziotis, T. Athanasiadis, I. Goudas, and J. Wipper. Mesh smoothing using the geometric element transformation method. *Computer Methods in Applied Mechanics and Engineering*, 197(45–48):3760–3767, 2008.

[144] D. Vartziotis and D. Bohnet. Regularizations of non-euclidean polygons. *ArXiv e-prints*, 1312.2500, 2013.

[145] D. Vartziotis and D. Bohnet. Convergence properties of a geometric mesh smoothing algorithm. *ArXiv e-prints*, 1411.3869, 2014.

[146] D. Vartziotis and D. Bohnet. Existence of an attractor for a geometric tetrahedron transformation. *Differential Geometry and its Applications*, 49:197–207, 2016.

[147] D. Vartziotis and D. Bohnet. A geometric mesh smoothing algorithm related to damped oscillations. *Computer Methods in Applied Mechanics and Engineering*, 326:102–121, 2017.

[148] D. Vartziotis, D. Bohnet, and B. Himpel. GETOpt mesh smoothing: Putting GETMe in the framework of global optimization-based schemes. *Finite Elements in Analysis and Design*, 147:10–20, 2018.

[149] D. Vartziotis and B. Himpel. Efficient and global optimization-based smoothing methods for mixed-volume meshes. In *Proceedings of the 22nd International Meshing Roundtable*, pages 293–311. Springer International Publishing, 2014.

[150] D. Vartziotis and B. Himpel. Efficient mesh optimization using the gradient flow of the mean volume. *SIAM Journal on Numerical Analysis*, 52(2):1050–1075, 2014.

[151] D. Vartziotis and S. Huggenberger. Iterative geometric triangle transformations. *Elemente der Mathematik*, 67(2):68–83, 2012.

[152] D. Vartziotis and J. Merger. GETMe.anis: On geometric polygon transformations leading to anisotropy. *ArXiv e-prints*, 1805.01767, 2018.

[153] D. Vartziotis and M. Papadrakakis. Improved GETMe by adaptive mesh smoothing. *Computer Assisted Methods in Engineering and Science*, 20(1):55–71, 2013.

[154] D. Vartziotis, A. Poulis, V. Fäßler, C. Vartziotis, and C. Kolios. Integrated Digital Engineering Methodology for Virtual Orthopedics Surgery Planning. In *Proceedings of the International Special Topic Conference on Information Technology in Biomedicine, Ioannina, Greece*, 2006.

[155] D. Vartziotis and J. Wipper. Classification of symmetry generating polygon-transformations and geometric prime algorithms. *Mathematica Pannonica*, 20(2):167–187, 2009.

[156] D. Vartziotis and J. Wipper. The geometric element transformation method for mixed mesh smoothing. *Engineering with Computers*, 25(3):287–301, 2009.

[157] D. Vartziotis and J. Wipper. On the construction of regular polygons and generalized Napoleon vertices. *Forum Geometricorum*, 9:213–223, 2009.

[158] D. Vartziotis and J. Wipper. Characteristic parameter sets and limits of circulant Hermitian polygon transformations. *Linear Algebra and its Applications*, 433(5):945–955, 2010.

[159] D. Vartziotis and J. Wipper. A dual element based geometric element transformation method for all-hexahedral mesh smoothing. *Computer Methods in Applied Mechanics and Engineering*, 200(9–12):1186–1203, 2011.

[160] D. Vartziotis and J. Wipper. Fast smoothing of mixed volume meshes based on the effective geometric element transformation method. *Computer Methods in Applied Mechanics and Engineering*, 201–204:65–81, 2012.

[161] D. Vartziotis and J. Wipper. The fractal nature of an approximate prime counting function. *Fractal and Fractional*, 1(1), 2017.

[162] D. Vartziotis, J. Wipper, and M. Papadrakakis. Improving mesh quality and finite element solution accuracy by GETMe smoothing in solving the Poisson equation. *Finite Elements in Analysis and Design*, 66:36–52, 2013.

[163] D. Vartziotis, J. Wipper, and B. Schwald. The geometric element transformation method for tetrahedral mesh smoothing. *Computer Methods in Applied Mechanics and Engineering*, 199(1–4):169–182, 2009.

[164] J. Vollmer, R. Mencl, and H. Müller. Improved Laplacian smoothing of noisy surface meshes. *Computer Graphics Forum*, 18(3):131–138, 1999.

[165] D. Wake, K. Lilja, and V. Moroz. A hybrid mesh generation method for two and three dimensional simulation of semiconductor processes and devices. In *Proceedings of the 7th International Meshing Roundtable*, pages 159–166, 1998.

[166] D.F. Watson. Computing the n-dimensional Delaunay tessellation with application to Voronoi polytopes. *The Computer Journal*, 24(2):167–172, 1981.

[167] J. Wipper. Mediale Achsen und Voronoj-Diagramme in der euklidischen Ebene. Master's thesis, Mathematisches Institut A, Universität Stuttgart, 1997.

[168] J. Wipper. *Finite-Elemente-Approximation mit WEB-Splines*. Shaker Verlag, 2005.

[169] T. Xia and E. Shaffer. Streaming tetrahedral mesh optimization. In *Proceedings of the 2008 ACM symposium on solid and physical modeling*, pages 281–286, 2008.

[170] H. Xu and T.S. Newman. An angle-based optimization approach for 2D finite element mesh smoothing. *Finite Elements in Analysis and Design*, 42(13):1150–1164, 2006.

[171] M.A. Yerry and M.S. Shephard. A modified quadtree approach to finite element mesh generation. *IEEE Computer Graphics and Applications*, 3(1):39–46, 1983.

[172] M.A. Yerry and M.S. Shephard. Automatic mesh generation for three-dimensional solids. *Computers & Structures*, 20(1–3):31–39, 1985.

[173] A.E. Yilmaz and M. Kuzuoglu. A particle swarm optimization approach for hexahedral mesh smoothing. *International Journal for Numerical Methods in Fluids*, 60(1):55–78, 2009.

[174] Y. Zhang and C. Bajaj. Adaptive and quality quadrilateral/hexahedral meshing from volumetric data. In *Proceedings of the 13th International Meshing Roundtable*, pages 365–376, 2004.

[175] Y. Zhang, C. Bajaj, and G. Xu. Surface smoothing and quality improvement of quadrilateral/hexahedral meshes with geometric flow. *Communications in Numerical Methods in Engineering*, 25(1):1–18, 2007.

[176] Y. Zhang, T.J.R. Hughes, and C.L. Bajaj. An automatic 3D mesh generation method for domains with multiple materials. *Computer Methods in Applied Mechanics and Engineering*, 199(5–8):405–415, 2010.

[177] M. Zhihong, M. Lizhuang, Z. Mingxi, and L. Zhong. A modified Laplacian smoothing approach with mesh saliency. In A. Butz, B. Fisher, A. Krüger, and P. Olivier, editors, *Smart Graphics*, volume 4073 of *Lecture Notes in Computer Science*, pages 105–113. Springer, 2006.

[178] T. Zhou and K. Shimada. An angle-based approach to two-dimensional mesh smoothing. In *Proceedings of the 9th International Meshing Roundtable*, pages 373–384, 2000.

[179] J.Z. Zhu, O.C. Zienkiewicz, E. Hinton, and J. Wu. A new approach to the development of automatic quadrilateral mesh generation. *International Journal for Numerical Methods in Engineering*, 32(4):849–866, 1991.

[180] O.C. Zienkiewicz, R.L. Taylor, and J.Z. Zhu. *The Finite Element Method: Its Basis and Fundamentals*. Butterworth-Heinemann, 7th edition, 2013.

Index